W9-ACG-352

Innovation Policy and the Economy 1

Innovation Policy and the Economy 1

edited by
Adam B. Jaffe, Josh Lerner, and Scott Stern

National Bureau of Economic Research
Cambridge, Massachusetts

The MIT Press
Cambridge, Massachusetts
London, England

F.W. Olin College Library

NBER/*Innovation Policy and the Economy*, Number 1, 2000
ISSN: 1531-3468
ISBN: Hardcover 0-262-10088-6
Paperback 0-262-60041-2
Published annually by The MIT Press, Cambridge, Massachusetts 02142

An electronic, full-text version of NBER/*Innovation Policy and the Economy* is available from MIT Press Journals when purchasing a subscription.

Subscription Rates
Hardcover/Print and Electronic: $58.00
Paperback/Print and Electronic: $24.00
Outside the U.S. and Canada add $10.00 for postage and handling. Canadians add 7% GST.

Subscription and address changes should be addressed to:
MIT Press Journals, Five Cambridge Center, Cambridge, MA 02142-1407, phone 617-253-2889; fax 617-577-1545; email: journals-orders@mit.edu. Claims will be honored free of charge if made within three months of the publication date of the issue. Claims may be submitted to journals-claims@mit.edu. Prices subject to change without notice.

In the United Kingdom, continental Europe, and the Middle East and Africa, send back volume orders and business correspondence to:
The MIT Press, Ltd., Fitzroy House, 11 Chenies Street, London WC1E 7ET England, phone 44-020-7306-0603, fax 44-020-7306-0604, email info@hup-MITpress.co.uk

In the United States and for all other countries, send single copy and back volume orders to:
The MIT Press, Five Cambridge Center, Cambridge, MA 02142, toll-free book orders 800-356-0343, fax 617-625-6660, email mitpress-orders@mit.edu

Copyright Information
Permission to photocopy articles for internal or personal use, or the internal or personal use of specific clients, is granted by the copyright owner for users registered with the Copyright Clearance Center (CCC) Transactional Reporting Service, provided that the fee of $10.00 per copy is paid directly to CCC, 222 Rosewood Drive, Danvers, MA 01923. The fee code for users of the Transactional Reporting Service is: 1531-3468/00 $10.00. For those organizations that have been granted a photocopy license with CCC, a separate system of payment has been arranged.

© 2001 by the National Bureau of Economic Research and the Massachusetts Institute of Technology.

This book was set in Palatino by Wellington Graphics, Westwood, MA.

Printed and bound in the United States of America.

National Bureau of Economic Research

Officers

Carl F. Christ, *Chairman*
Kathleen B. Cooper, *Vice Chairman*
Martin Feldstein, *President and Chief Executive Officer*
Robert Mednick, *Treasurer*
Susan Colligan, *Corporate Secretary*
Kelly Horak, *Controller and Assistant Corporate Secretary*
Gerardine Johnson, *Assistant Corporate Secretary*

Directors at Large

Peter C. Aldrich
Elizabeth E. Bailey
John H. Biggs
Andrew Brimmer
Carl F. Christ
Don R. Conlan
Kathleen B. Cooper
George C. Eads
Martin Feldstein
Stephen Friedman
George Hatsopoulos
Karen N. Horn
Judy C. Lewent
John Lipsky
Michael H. Moskow
Alicia H. Munnell
Rudolph A. Oswald
Robert T. Parry
Peter G. Peterson
Richard N. Rosett
Kathleen P. Utgoff
Marina v. N. Whitman
Martin B. Zimmerman

Directors by University Appointment

George Akerlof, *California, Berkeley*
Jagdish Bhagwati, *Columbia*
William C. Brainard, *Yale*
Glen G. Cain, *Wisconsin*
Franklin Fisher, *Massachusetts Institute of Technology*
Saul H. Hymans, *Michigan*
Marjorie B. McElroy, *Duke*
Joel Mokyr, *Northwestern*
Andrew Postlewaite, *Pennsylvania*
Nathan Rosenberg, *Stanford*
Michael Rothschild, *Princeton*
Craig Swan, *Minnesota*
David B. Yoffie, *Harvard*
Arnold Zellner, *Chicago*

Directors by Appointment of Other Organizations

Mark Drabenstott, *American Agricultural Economics Association*
Gail D. Fosler, *The Conference Board*
A. Ronald Gallant, *American Statistical Association*
Robert S. Hamada, *American Finance Association*
Robert Mednick, *American Institute of Certified Public Accountants*
Angelo Melino, *Canadian Economics Association*
Richard D. Rippe, *National Association For Business Economics*
John J. Siegfried, *American Economic Association*

David A. Smith, *American Federation of Labor and Congress of Industrial Organizations*
Josh S. Weston, *Committee for Economic Development*
Gavin Wright, *Economic History Association*

Directors Emeriti

Moses Abramovitz
Thomas D. Flynn
Lawrence R. Klein
Franklin A. Lindsay
Paul W. McCracken
Bert Seidman
Eli Shapiro

Since this volume is a record of conference proceedings, it has been exempted from the rules governing critical review of manuscripts by the Board of Directors of the National Bureau (resolution adopted 8 June 1948, as revised 21 November 1949 and 20 April 1968).

Contents

Introduction   ix
Adam B. Jaffe, Josh Lerner, and Scott Stern

1 **Publicly Funded Science and the Productivity of the
Pharmaceutical Industry**   1
Iain M. Cockburn and Rebecca M. Henderson

2 **Creating Markets for New Vaccines—Part I: Rationale**   35
Michael Kremer

3 **Creating Markets for New Vaccines—Part II: Design Issues**   73
Michael Kremer

4 **Navigating the Patent Thicket: Cross Licenses, Patent Pools, and
Standard Setting**   119
Carl Shapiro

5 **Commercialization of the Internet: The Interaction of Public
Policy and Private Choices or Why Introducing the Market
Worked So Well**   151
Shane Greenstein

6 **Numbers, Quality, and Entry: How Has the Bayh-Dole Act
Affected U.S. University Patenting and Licensing?**   187
David C. Mowery and Arvids A. Ziedonis

7 **Should the Government Subsidize Supply or Demand in the
Market for Scientists and Engineers?**   221
Paul M. Romer

Introduction

This volume is the inaugural publication of the National Bureau of Economic Research (NBER) Innovation Policy and the Economy (IPE) group. The past few years have seen an increasing appreciation of the importance of innovation to the economy. The importance of innovation can be seen, for instance, in the market capitalization of technology-based companies, the number of patent filings by U.S. corporations, or the coverage of new innovations in the business press.

At the same time, there is active debate regarding the implications of rapid technological change for economic policy, and the appropriate policies and programs regarding research, innovation, and the commercialization of new technology. These debates encompass long-standing issues, such as the appropriate level and form of public support of research, as well as newer issues, such as the evolving role of intellectual property and the appropriate antitrust treatment of software and other industries where technology standards play a key role.

The IPE group seeks to provide an accessible forum to bring the work of leading academic researchers to an audience of policymakers and those interested in the role of public policy and innovation. Our goals are: (1) to provide an ongoing forum for the presentation of research on the impact of public policy on the innovative process; (2) to stimulate such research by exposing potentially interested researchers to the issues that policymakers consider important; and (3) to increase the awareness of policymakers (and the public policy community more generally) concerning contemporary research in economics and the other social sciences that usefully informs the evaluation of current or prospective proposals relating to innovation policy.

This volume contains the papers presented at the group's first meeting, held in Washington DC in April 2000. Subsequent volumes will contain the proceedings of annual meetings in Washington each spring.

In addition, the group meets annually in Cambridge for discussion of research in progress, and will hold occasional conferences on specific topics.

The papers in this volume demonstrate the importance of issues related to innovation in current policy debates, the value of the insights that economic analysis can bring to these problems, and the breadth of interest of economics in innovation-related issues. The first two papers highlight the interaction between public policy and innovation in a specific but important sector—the life sciences.

Motivated by the extraordinary rise in public expenditures on the life sciences over the past 30 years, Iain Cockburn and Rebecca Henderson initiate the volume by assessing the relationship between public investment in life sciences research and the rate and extent of innovation in the pharmaceutical industry. Though the social benefits provided by basic research are notoriously difficult to quantify, economic research on the industry has identified several specific mechanisms through which public funding may spur innovation and the commercialization of new therapies. For example, public funding of basic research in molecular biology has provided critical elements of the foundation of rational drug design—a more efficient drug discovery technique whereby researchers investigating new compounds are guided by scientific evidence about the biochemical basis of disease. Building both on case histories of specific drugs as well as evidence based on patenting and publication data from the industry, the paper assesses the returns to public expenditure in this sector. While noting the limitations and assumptions associated with individual studies, their review suggests that prior econometric research "makes a quite convincing case for a high rate of return to public science."

Michael Kremer's paper, on the other hand, examines an arena where private pharmaceutical companies do very little research: the development of vaccines for tropical diseases. He argues that the reluctance of pharmaceutical companies to undertake research to develop vaccines for diseases such as malaria, tuberculosis, and African strains of HIV is a consequence of severe market failures. In particular, companies fear that were they to develop such products, governments would force prices down to a level that would not allow them to earn a satisfactory return. To address this problem, he proposes that public agencies commit to buying vaccines at a set price if a satisfactory vaccine can be developed by the private sector. Kremer suggests that such an initiative could address market failures far more effectively than alternative approaches, such as government subsidies for basic research in this area.

Carl Shapiro focuses on the increasingly important issue of the interaction between intellectual property protection and competition policy. He notes that in several important technology areas, such as biotechnology, semiconductors, and software, commercial innovation often requires use of numerous potentially overlapping or conflicting patent rights. A particularly important circumstance in which such problems often arise is where standard setting is an essential part of the process by which new technologies are commercialized. Parties can use a variety of contractual mechanisms to resolve these problems, including cross licensing, package licensing, and patent pools. Such agreements, however, are sometimes challenged by antitrust authorities, particularly when they occur between or among horizontal competitors. Shapiro offers several suggestions as to how agreements that facilitate innovation can be distinguished from those with serious anticompetitive effects.

Shane Greenstein's paper asks the question: why did the commercialization of the Internet go so well? This question is intriguing and also important, because most historical examples of technology transfer from the public to the private sector have involved confusion and delay. In contrast, the technologies that make up the Internet diffused rapidly and pervasively shortly after the National Science Foundation relaxed the regulations restricting private use in 1992. Drawing upon his own research on Internet Service Providers, Greenstein identifies four drivers of this unusually quick diffusion process: (1) the absence of significant technical or commercial hurdles; (2) the economic and technical malleability of the Internet; (3) the potential for customization on a number of key dimensions; and (4) the fortuitous coincidence that the Internet was commercialized at the same time that the World Wide Web was developed. The diffusion of the Internet yields two policy insights. First, the Internet experience may allow for better identification of the conditions under which technology may be transferred successfully from the public to the private sector. Second, the Internet has had a ubiquitous impact on the telecommunications industry in particular, perhaps necessitating a rethinking of the regulatory institutions underlying this sector of the economy.

The last two papers in the volume examine the different facets of the relationship between academic institutions and innovation. The first of these, by David Mowery and Arvids Ziedonis, examines a direct impact: the commercialization of academic discoveries by the private sector. Drawing on in-depth studies of three leading research universities—Columbia, Stanford, and the University of California—the authors examine the impact of the policy reforms of the early 1980s on technology transfer activities. Disentangling the impact of this policy change from the contemporaneous shifts in federal funding of research and patent policy is challenging. But the results suggest that, at least at these three schools, the reforms of federal technology transfer policy served as a significant boost to commercial activities. This suggests that the Bayh-Dole Act of 1980 is helping to achieve its stated goal of increasing the transfer of commercially useful technology from universities to the private sector.

Although commercialization of university technology is important, most scholars agree that the primary contribution of academia to commercial innovation is its training of scientists and engineers. In the last paper of this volume, Paul Romer notes that, in principle, government policy to foster innovation could act on both the demand and supply sides of the R&D investment process. He argues that, historically, policy has focused more on the demand side, using instruments such as R&D tax credits or other subsidies. If the supply of R&D resources—primarily technically trained people—adjust only very slowly to changes in policy, then such policies will raise the wages of scientists and engineers without increasing research much. Romer then analyzes how the structure and incentives of post-secondary and graduate education in the U.S. affect the supply of technically trained people to the commercial sector. He concludes that undergraduate education discourages students from majoring in science or engineering, and that post-graduate programs are structured to produce Ph.D.s oriented toward teaching rather than commercial research. He then considers several government policies that might counter these tendencies, thereby increasing the supply of scientists and engineers and ultimately the rate of commercial innovation and economic growth.

These six essays highlight the role that economic theory and empirical analysis can play in evaluating key policies impacting innovation. Together, they offer the prospect that contemporary research in economics can usefully inform the evaluation of current and prospective innovation policy alternatives.

As a final note, we gratefully acknowledge several key actors who have helped to launch the IPE group and this inaugural volume. Kirsten Davis and Rob Shannon of the NBER Conference Department provided critical support in organizing the April 2000 conference. Helena Fitz-Patrick of the NBER and Janet Fisher of the MIT Press have shepherded the papers from that conference into this volume. This initiative was part of the NBER Project on Industrial Technology and Productivity, with support of the Alfred P. Sloan Foundation. Ernst Berndt and the late Zvi Griliches, as directors of the NBER Productivity Program, were supportive of the effort since its inception. Finally, Martin Feldstein, President and CEO of the NBER, generously provided encouragement and support for this new venture.

Adam B. Jaffe, Josh Lerner, and Scott Stern

1

Publicly Funded Science and the Productivity of the Pharmaceutical Industry

Iain M. Cockburn, *Boston University and NBER*
Rebecca M. Henderson, *MIT and NBER*

Executive Summary

U.S. taxpayers funded $14.8 billion of health related research last year, four times the amount that was spent in 1970 in real terms. In this paper we evaluate the impact of these huge expenditures on the technological performance of the pharmaceutical industry. While it is very difficult to be precise about the payoffs from publicly funded research, we conclude from a survey of a wide variety of quantitative and qualitative academic studies that the returns from this investment have been large, and may be growing even larger. Public sector science creates new knowledge and new tools, and produces large numbers of highly trained researchers, all of which are a direct and important input to private sector research. But this is not a one way street: the downstream industry is closely linked with upstream institutions, and knowledge, materials, and people flow in both directions. One important contribution of public science is that it sustains an environment in which for-profit firms can conduct their own basic research, which in turn contributes to the global pool of knowledge. Measured quite narrowly in terms of its effect on private sector R&D, the rate of return to public funding of biomedical sciences may be as high as 30% per year. Large as this figure is, these calculations are likely an underestimate, since they fail to fully capture the wider impact of pharmaceutical innovation on health and well-being. Indeed, the best may be yet to come: the revolution in molecular biology that began in publicly funded laboratories 25 years ago—and continues to be driven by the academic research—promises dramatic advances in the treatment of disease.

I. Introduction

Between 1970 and 1999, public funding in the U.S. for health related research increased over 400% in real terms, to $14.8 billion, or 38% of the non-defense Federal research budget. Worldwide, the U.S. spends more of its publicly available research funds on human health than any other nation (table 1.1). What kind of impact is this research having? Is

Table 1.1
National expenditures on academic and related research by main field, 1987

	Expenditure (1987 M$)[a]						
	U.K.	FRG	France	Neth.	U.S.	Japan	Average[b]
Engineering	436	505	359	112	1,966	809	14.3%
	15.6%	12.5%	11.2%	11.7%	13.2%	21.6%	
Physical	565	1,015	955	208	2,325	543	21.2%
Sciences	20.2%	25.1%	29.7%	21.7%	15.6%	14.5%	
Life Sciences	864	1,483	1,116	313	7,285	1,261	36.3%
	30.9%	36.7%	34.7%	32.7%	48.9%	33.7%	
Social Sciences	187	210	146	99	754	145	6.0%
	6.7%	5.2%	4.6%	10.4%	5.1%	3.9%	
Arts &	184	251	218	83	411	358	6.8%
Humanities	6.6%	6.2%	6.8%	8.6%	2.8%	9.6%	
Other	562	573	418	143	2,163	620	15.6%
	20.1%	14.2%	13.0%	14.9%	14.5%	16.6%	
Total	2,798	4,037	3,212	958	14,904	3,736	

[a]Expenditure data are based on OECD "purchasing power parities" for 1987 calculated in early 1989.
[b]This represents an unweighted average for the six countries (i.e., national figures have not been weighted to take into account the differing size of countries).
Irvine, J., B. Martin, and P. Isard 1990 p. 219.

the public getting an appropriate "bang for its buck"? This paper explores one aspect of this question by focusing on one issue in particular: the impact of publicly funded research on the productivity of the U.S. pharmaceutical industry.

The pharmaceutical industry provides a particularly interesting window through which to study the more general question of the impact of publicly funded research. The public sector probably plays a more important role in determining private sector productivity in the pharmaceutical industry than in any other industry except defense. Public sector research spending almost equals private sector spending, and publicly funded researchers generate a disproportionate share of the papers published in the relevant fields (Stephan 1996). It is also the case that scientific advances in medical practice appear to have had a very significant effect on human health. Between 1940 and 1990, average life expectancy in the U.S. increased from 63.6 to 75.1 years, and the average quality of life appears to have also increased over the same time period (Cutler and Richardson 1999). While advances in human health have many causes, advances in pharmaceutical therapies have made a very significant contribution. New drugs have revolutionized the treatment of ulcers, stroke, and various psychiatric conditions. They have dramatically improved the quality of life of asthma sufferers. They have brought the symptoms of AIDS under control for a significant fraction of the infected population. Some cancers are now reliably curable by drug therapy, and new drugs for hypertension and high cholesterol are proving instrumental in the treatment of heart disease, still the largest killer of Americans. Drugs "in the pipeline" promise major advances in the treatment of arthritis, Alzheimer's disease, many kinds of cancers and a variety of other chronic conditions. Since there is general agreement among qualified observers that publicly funded science has played a major role in these advances, the industry presents a particularly salient setting in which to explore its economic impact.

This paper begins by briefly reviewing the progress that has been made in estimating the rate of return to publicly funded science in other settings. Efforts to measure the rate of return to public research in any context are dogged by a variety of difficult practical and conceptual problems (Griliches 1979; Jones and Williams 1995), and many of these problems are particularly severe in the case of the pharmaceutical industry. We outline some of the difficulties inherent in generating quantitative estimates, and summarize some key results. Although measuring the research output of the public sector and its impact on the rest of the economy presents enormous challenges, both quantitative and qualitative estimates suggest that the rate of return to basic research in general is probably quite high. Case studies of specific technologies and government programs point to the critical role of public sector research in laying the foundation for technological advances that have later had enormous impact on the civilian economy (see for example David, Mowery, and Steinmuller 1992), and direct quantitative estimates suggest that the rate of return to publicly funded research is on the order of 25–40%, (Adams 1990; Mansfield 1991; Griliches 1979, 1994).

The paper then turns to a discussion of the pharmaceutical industry. We show that publicly funded research has increased dramatically over the last 60 years, and present some evidence suggesting that its role has become increasingly important. While pharmaceutical research in the 1960s and 1970s drew upon the results of federally funded research, it typically took many years for the results of publicly funded work to have an impact on the private sector, and many firms made only limited use of publicly generated results. The revolution in molecular biology and the transition to "rational" or "mechanism driven" drug design and then to the techniques of biotechnology have revolution-ized the relationship between the public and private sectors, making immediate access to leading edge publicly funded science a key com-petitive advantage for leading pharmaceutical firms and stimulating the development of an entirely new segment of the industry: the small biotechnology firm. While early research in the industry followed a more traditional waterfall model, with the results of publicly funded research gradually flowing downhill to the private sector, over the last 25 years the relationship has become much more that of an equal part-nership, with ideas and materials flowing upstream to the public sector as well as downstream to industry.

This is followed by a discussion of the quantitative evidence. We show that many of the problems that make the measurement of the im-pact of publicly funded research difficult in the wider economy operate with particular force here: it is quite difficult to measure either the "output" of publicly funded research or its impact on the private sec-tor, there are long and variable lags between the generation of knowl-edge in the public sector and its impact in industry, and there are many different pathways through which the public sector shapes private sec-tor research. The public sector generates more than just scientific pa-pers, pure knowledge, and highly trained graduates: its existence also supports a community of "open science" that sustains high quality pri-vately funded research in the for-profit laboratories.

While it is, therefore, very hard to precisely estimate the return to publicly funded biomedical research, we nonetheless conclude that it is quite high. On the one hand, the qualitative evidence is compelling: the U.S. pharmaceutical and biotechnology industries lead the world, and while there are a variety of plausible reasons for this, the strength of the public research base is surely among the most important. Detailed case studies have highlighted the role of the public sector in supporting the development of important new drugs, and almost universal agreement among private sector researchers that publicly funded research is vital to all that they do. There are also a number of econometric studies that, while imperfect and undoubtedly subject to improvement and revi-sion, between them make a quite convincing case for a high rate of re-turn to public science in this industry. It is worth noting that there are, so far as we are aware, *no* systematic quantitative studies that have found a negative impact of public science!

We conclude the paper with a brief summary and some speculations as to the future. The ongoing revolution in genetics, genomics, and bioinformatics—all advances that have their roots squarely in federally funded research—promises to revolutionize the treatment of many diseases. If this promise is realized, the role of publicly funded research in advancing human health through the support of pharmaceutical innovation will be beyond question. This is an exciting period in which to be studying the impact of publicly funded research.

II.   Measuring the Impact of Publicly Funded Research: The General Problem

Government funding for "basic" or "fundamental" research has traditionally been justified on the grounds that the *social* returns to basic research are likely to significantly exceed *private* returns, and thus that the private sector will underinvest in basic research relative to the social optimum (Nelson 1959; Arrow 1962). Private firms are unlikely to be able to capture (or "appropriate") the returns to basic research because, in general, it usually takes a very long time for the practical implications of basic research to become apparent and because these implications are often highly diffuse. A firm that is funding basic research in optics, for example, might not see any return on its investment for many years, and many of these returns might be realized by its competitors or by firms competing in entirely different industries. There is also some evidence that basic research has its greatest impact when it is funded by the public sector because publicly funded researchers are more likely to compete for prestige in their fields than for financial gain, and since prestige is gained through the rapid publication of their results, publicly funded research is likely to become more rapidly available across the economy than privately funded work (Dasgupta and David 1987, 1994; Merton 1973).

Unfortunately exactly the same characteristics that make it economically desirable for the government to fund basic research—the long lags between research and impact and its wide diffusion across firms and industries—also make it very difficult to measure its effects. Two techniques have been widely used: case studies and econometric production function based analysis. The available case studies are suggestive, and most of them suggest that government funded research has indeed had a significant effect on private sector productivity (see, for example, David, Mowery, and Steinmuller 1992; Mansfield 1991; and the NSF 1968), but it is always difficult to know how far one can generalize from the results of case study research. Broad-based statistical estimates of the relationships between upstream funding and downstream performance could place this case study evidence on sounder footing, but these too have their problems.

III. Productivity Measurement

Productivity is a natural way to assess the performance of firms and industries. High levels of productivity and rapid growth in productivity are unambiguous indicators of the technological progress which public sector research is intended to support. Productivity relates output to input, and publicly funded research affects this relationship in a number of ways. Research results, experimental materials, and the human capital of highly trained researchers are "free" inputs to the production of drugs. At the same time basic research improves the efficiency with which these inputs are utilized as it identifies productive areas for investigation or provides new, more effective research tools.

There are good reasons to expect to see an important impact of publicly funded research in the productivity statistics for the pharmaceutical industry. The problem is that productivity measurement is beset by a number of difficult problems, which are exacerbated in knowledge-intensive industries with a high rate of new product development. To frame our subsequent discussion, we first review these difficulties.

Measuring Output

Productivity is the ratio of inputs to output.[1] We begin with the numerator. Measuring output is difficult in any research-driven industry for two closely related reasons. In the first place, output is conventionally measured by the sales of the level of the firm that produces it. There are good reasons to believe that this underestimates the social value of output. If there are significant externalities, or "spillovers," these will not be captured in the price charged to consumers. Vaccines provide a good example: purchasing the product provides private benefits to the individual who is vaccinated, but also society benefits from the impact that vaccination has on lowering the prevalence of the disease—or in the case of smallpox, wiping it out altogether.

The second problem is that research tends to improve the quality of output, and quality improvement may or may not be reflected in the price of output. The computer industry provides the classic example of this problem: the price of a standard PC has remained more or less constant over the last 5 years, but the power and capabilities of the standard package has consistently increased over time. Statistical methods for addressing this problem have revealed that the "true" quality- adjusted price of a PC has been falling at a very rapid rate: 20% per year or more. The same may be true for pharmaceutical products. New drugs tend to be more effective, have fewer side effects, and can be taken in much more convenient dosages (once per day instead of four times per day). But efforts to apply quality adjustments to reflect these benefits to the consumer have had limited success. While the quality attributes of a car or a PC are relatively straightforward to measure (fuel efficiency, weight, processor speed, storage capacity, etc.) those of pharmaceutical products are more difficult to define and measure. (See for example Berndt et al. 2000; and Cockburn and Anis 2000.)

For pharmaceuticals, an additional complication is introduced by the tangled web of economic relationships between patients, physicians, pharmacists, and insurers. In some industries, prices are a reasonably good measure of consumers' willingness to pay. Here, the link between expenditures on drugs and the value to the consumer is harder to draw. In some cases, the value of a drug to the patient who receives it may be considerably greater than the price received by the pharmaceutical company which produces it. (It is also possible that some pharmaceuticals are worth less to patients than the price that is charged for them.)

Some researchers have used measures of output such as quality adjusted life years (QUALYs) to correct for these problems, but any such attempt must inevitably rest on a series of assumptions and judgments that will always be open to question (Cutler and Richardson 1999).

Measuring Inputs

Measuring inputs to the production of pharmaceuticals raises similar problems. To produce pharmaceuticals requires not just person hours, capital, energy, materials and so forth, but also "knowledge capital" or know-how. Measuring knowledge capital is particularly difficult since the impact of knowledge is felt over an extended period of time. Labor generates output today: but scientific explorations often only produce tangible output after many years. For example in a study of the impact of scientific research across the entire economy, Adams (1990) found evidence that *on average* it takes 20 years for basic research to produce tangible economic results! In the context of the pharmaceutical industry this problem is particularly acute. Not only is it the case that the lags between the publication of any particular piece of publicly funded research and its impact on the discovery of a new drug are long and highly variable, but on average it takes 12–14 years to translate any given private sector discovery into a drug that can be given to patients. This problem is particularly concrete in the case of federal funding for molecular biology in general and genetics in particular. While this funding has had some immediate impact on industry productivity, there is widespread (even universal) agreement that in this case the best is yet to come. The interpretation of the human genome, for example, is not expected to have a direct effect on human health for many years.

The measurement of knowledge inputs in pharmaceuticals—and of the effect of publicly funded research in particular—is further complicated by the fact that there are a multitude of mechanisms through which publicly funded research shapes and supports private sector productivity. The most straightforward is through the development of new scientific knowledge. Figure 1.1 presents a much simplified diagram of the process of drug discovery and development.

In general, fundamental research (the discovery of fundamental scientific knowledge) precedes "drug discovery" (the search for compounds that seem to work in test tubes and/or in animals), which is followed by "drug development" (the process by which one makes sure that seemingly useful compounds actually work safely in humans). But while it is certainly the case that publicly funded research has been responsible for generating an enormous amount of fundamental science that has supported major breakthroughs in the industry, this is by no means the only way in which public investment in biomedical research has shaped the industry.

The public sector supports private sector research through a variety of other mechanisms. One of the most important of these is the provision of trained scientists, but the impact of others should not be overlooked: public sector science also includes discovery of new research tools, direct investment in a small amount of drug discovery work, and funding of leading edge work in clinical development.

Studies of the relationship between universities and the private sector in general have suggested that one of the most important outputs of

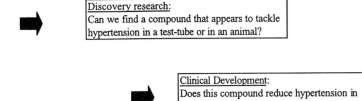

Figure 1.1
A simple model of drug discovery

the university sector is trained personnel (see for example Agarwal 2000). This is likely to be particularly true in the pharmaceutical industry, where research is conducted by battalions of skilled scientists, many of them with doctorates and postgraduate educations funded in large part through NSF and NIH grants. As industry hires these graduates it benefits not only from their general training and skills but also from the leading edge access it gives them to research being conducted within the public sector. Another important publicly funded input to the private sector is not new knowledge, per se (how do viruses metabolize?) but new *tools*. Cohen and Boyer's discovery of one of the most fundamental tools of genetic engineering is one prominent example of this phenomenon.

Figure 1.1 suggests that the process of drug discovery is an almost entirely linear one, with scientific knowledge feeding directly into drug discovery. In reality, however, the interaction between the public and private sectors is much more iterative and complex. In *Networks of Innovation*, for example, Galambos and Sewell (1995) show that the development of vaccines was characterized by the continual exchange of information between researchers working at Merck and researchers working in the public sector. On several occasions the development of novel therapies by the private sector or advances made in clinical treatment have *preceded* major advances in fundamental knowledge. Brown and Goldstein's Nobel Prize winning work on the structure of the LDL receptor, for example, occurred simultaneously with the discovery of the first effective HMG CoAse reductase inhibitors; and the recognition that stomach ulcers are bacterial in origin flowed from the pioneering work of physicians working in the clinic rather than from basic scientific research. Even apparently straightforward cases such as the discovery of AZT appear, on closer examination, to have a fine grained structure that reflects a bidirectional flow of knowledge rather than the simple transmission of research results or new ideas from the public to private sectors (Cockburn and Henderson 1998).

This bidirectional structure not only complicates the problem of imposing a time structure on the estimation of the effect of public research (see below). It also hints at another important role of public research: the maintenance of a community of researchers, or a public rank hierarchy in which private sector researchers can be evaluated and promoted on the basis of their standing in the public community of science. As the techniques of drug discovery evolved and it became increasingly important to be able to take advantage of the findings of public science, the most productive pharmaceutical firms began to reward their researchers on the basis of their standing in the eyes of their peers (Henderson and Cockburn 1994; Cockburn, Henderson, and Stern 2000). To some degree the adoption of this incentive mechanism undoubtedly reflects the fact that it encourages a firm's scientist to publish and to engage with the community of public scientists, and this in turn facilitates the firm's ability to take advantage of publicly generated knowledge. But its adoption also probably solves a difficult problem for managers: evaluating the effort and performance of scientific professionals whose work is becoming increasingly complex and increasingly difficult to monitor. To the degree that these practices increase the flexibility and creativity of the private sector, the maintenance of a public community of science acts as an input (of a particularly subtle kind) to the private sector.

Arguably, the presence of the community of open science also provides an implicit subsidy to the industry in that it provides important nonmonetary rewards. Scott Stern (Stern 1999) has shown that researchers are willing to trade salary for the opportunity to work on scientifically interesting projects. Since salaries are a large fraction of total research costs, to the extent that this phenomenon drives down the wages demanded by scientists, the industry benefits.

The public sector also invests in the actual discovery of new drugs through its support of screening programs such as that conducted by the National Institute for Cancer. While this program has generated one important new drug—Taxol—its overall impact appears to be minimal. Last, but by no means least, the public sector supports private sector productivity through the support of clinical development and clinical research. There is some evidence that this type of research provides a critically important stimulus to the discovery of new drugs (Wurtman and Bettiker 1994, 1996).

The measurement of knowledge capital is further complicated by the problem of "spillovers," or the fact that knowledge generated in one place or firm is often useful elsewhere. At the level of the entire economy this effect is unproblematic, but when one is trying to measure the impact of spillovers from publicly funded research on a particular firm, for example, it raises serious problems. Where should one look for spillovers? One important source may be a firm's competitors. In the case of pharmaceuticals, for example, we showed that private sector research productivity was directly and significantly affected by competitive research activity (Henderson and Cockburn 1996). Then there is the question of whether to treat all federally funded entities equally, or to trace spillovers only to those that are geographically or technically "close." And while the U.S. government accounts for a substantial fraction of worldwide public sector research, science is a global enterprise. Contributions from significant publicly funded research activity in Europe and elsewhere ought not to be ignored.

Despite these problems, a number of researchers have attempted to use productivity measures to estimate the rate of return to publicly funded research. Studies at the aggregate level, or at the level of the entire economy, generate numbers in the 20–40% range, as described above. As an illustration of these results, table 1.2 reproduced from Griliches 1995, summarizes the results from a number of studies of industry productivity. The results vary widely, but seem to suggest that the rate of return to public research is likely to be quite high.[2]

IV.   The Role of Publicly Funded Research in the Pharmaceutical Industry

Attempts to measure the role of publicly funded research in the context of the pharmaceutical industry must not only grapple with these issues, but must also take account of an environment in which the relationship between the public and private sector has changed dramatically over the last 50 years.

Public funding for health related research is largely a product of the Second World War. Before the war the pharmaceutical industry was

Table 1.2
Selected estimates of returns to R&D and R&D spillovers

Agriculture	Rate of return to public R&D
Griliches (1958) Hybrid corn	35–40
Hybrid sorghum	20
Peterson (1967) Poultry	21–25
Schmitz-Secker (1970) Tomato harvester	37–46
Griliches (1964) Aggregate	35–40
Evenson (1968) Aggregate	41–50
Knutson-Tweeten (1979) Aggregate	28–47
Huffman-Evenson (1993) Crops	45–62
Livestock	11–83
Aggregate	43–67
Industry	**Rate of return to all R&D**
Case studies	
Mansfield et al. (1977)	25–56
I-O Weighted	
Terleckyj (1974): Total	28–48
Private	29–78
Sveikausakas (1981)	10–50
Goto-Suzuki (1989)	26–80
R&D Weighted (patent flows)	
Griliches-Lichtenberg (1984)	46–69
Mohnen-Lepine (1988)	28–56
Cost Functions	
Bernstein-Nadiri (1988, 1989)	
Differs by industry	9–160
Bernstein-Nadiri (1991)	14–28

Table 3.4 from Griliches, 1995.

not tightly linked to formal science. Until the 1930s, when sulfonamide was discovered, drug companies undertook little formal research. Most new drugs were based on existing organic chemicals or were derived from natural sources (e.g., herbs) and little formal testing was done to ensure either safety or efficacy. Harold Clymer, who joined SmithKline (a major American pharmaceutical company) in 1939, noted:

[Y]ou can judge the magnitude of [SmithKline's] R&D at that time by the fact I was told I would have to consider the position temporary since they had already hired two people within the previous year for their laboratory and were not sure that the business would warrant the continued expenditure. (Clymer, 1975)

World War II and wartime needs for antibiotics marked the drug industry's transition to an R&D intensive business. Penicillin and its antibiotic properties were discovered by Alexander Fleming in 1928.

However, throughout the 1930s, it was produced only in laboratory scale quantities and was used almost exclusively for experimental purposes. With the outbreak of World War II, the U.S. government organized a massive research and production effort that focused on commercial production techniques and chemical structure analysis. More than 20 companies, several universities, and the Department of Agriculture took part.

The commercialization of penicillin marked a watershed in the industry's development. Due partially to the technical experience and organizational capabilities accumulated through the intense wartime effort to develop penicillin, as well as to the recognition that drug development could be highly profitable, pharmaceutical companies embarked on a period of massive investment in R&D and built large-scale internal R&D capabilities. At the same time there was a very significant shift in the institutional structure surrounding the industry. Whereas before the war public support for health related research had been quite modest, after the war it boomed to unprecedented levels. The period from 1950 to 1990 was a golden age for the pharmaceutical industry, as for industry in general, and particularly the major U.S. players—firms such as Merck, Eli Lilly, Bristol-Myers, and Pfizer—grew rapidly and profitably. R&D spending exploded and with this came a steady flow of new drugs. (Figure 1.2 shows publicly funded spending on health related research and total U.S. R&D spending by U.S. pharmaceutical firms.[3])

A number of factors supported the industry's high level of innovation. One was the sheer magnitude of both the research opportunities and the unmet needs. In the early postwar years, there were many diseases for which no drugs existed. In every major therapeutic category—from painkillers and anti-inflammatories to cardiovascular and central nervous system products—pharmaceutical companies faced an almost completely open field (before the discovery of penicillin, very few drugs effectively *cured* diseases).

Faced with such a target rich environment but very little detailed knowledge of the biological underpinnings of specific diseases, pharmaceutical companies invented an approach to research now referred to as "random screening." Under this approach, natural and chemically derived compounds are randomly screened in test tube experiments and laboratory animals for potential therapeutic activity. Pharmaceutical companies maintained enormous libraries of chemical compounds, and added to their collections by searching for new compounds in places such as swamps, streams, and soil samples.

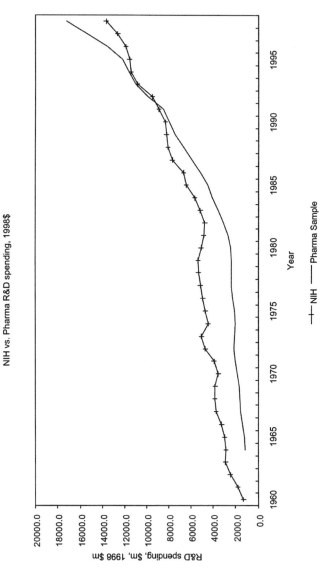

Figure 1.2
Changes in R&D spending over time

Thousands, if not tens of thousands, of compounds might be subjected to multiple screens before researchers honed in on a promising substance. Serendipity played a key role since in general the "mechanism of action" of most drugs—the specific biochemical and molecular pathways that were responsible for their therapeutic effect—were not well understood. Researchers were generally forced to rely on the use of animal models as screens. For example researchers injected compounds into hypertensive rats or dogs to explore the degree to which they reduced blood pressure. Under this regime it was not uncommon for companies to discover a drug to treat one disease while searching for a treatment for another. Although random screening may seem inefficient, it worked extremely well for many years, and continues to be widely employed. Several hundred chemical entities were brought to the market in the 1950s and 1960s and several important classes of drug were discovered in this way, including a number of important diuretics, all of the early vasodilators, and a number of centrally acting agents including reserpine and guanethidine.

In general, this early form of random screening made only delayed and indirect use of the results of publicly funded research. Beginning in the early 1970s, the industry began to benefit more directly from the explosion in public funding for health related research that followed the war. Publicly funded research had always been important to the industry's health, but initially it was probably most important as a source of knowledge about the etiology of disease. For example it was the publicly funded Framingham heart study that showed that elevated blood pressure (hypertension) was associated with a greater risk of heart disease and death, and thus encouraged the industry to search for drugs that might tackle it.

From the middle 1970s on, however, substantial advances in physiology, pharmacology, enzymology, and cell biology—the vast majority stemming from publicly funded research—led to enormous progress in the ability to understand the mechanism of action of some existing drugs and the biochemical and molecular roots of many diseases. This new knowledge made it possible to design significantly more sophisticated screens. By 1972, for example, the structure of the renin angiotensive cascade, one of the systems within the body responsible for the regulation of blood pressure, had been clarified by the work of Laragh and his collaborators (Laragh et al. 1972) and by 1975 several companies had drawn on this research in designing screens for hypertensive drugs (Henderson and Cockburn 1994). These firms could replace ranks of hypertensive rats with precisely defined chemical reactions. In place of the request "find me something that will lower blood pressure in rats" pharmacologists could make the request "find me something that inhibits the action of the angiotensin 2 converting enzyme."

The more sensitive screens in turn made it possible to screen a wider range of compounds. Prior to the late 1970s, for example, it was difficult to screen the natural products of fermentation (a potent source of new antibiotics) in whole animal models. The compounds were available in such small quantities, or triggered such complex mixtures of reactions in living animals, that it was difficult to evaluate their effectiveness. The use of enzyme systems as screens made it much easier to screen these kinds of compounds. It also triggered a virtual cycle in that the availability of drugs whose mechanisms of action were well known made possible significant advances in the medical understanding of the natural history of a number of key diseases, which in turn opened up new targets and opportunities for drug therapy.

Both "random" and "guided" or "science driven" drug discovery continue to be important tools in the search for new drugs,[4] but the most important development in the pharmaceutical industry is the advent of the science and techniques of biotechnology—and in this field the role of publicly funded research is even more pronounced. Historically, most drugs have been derived from natural sources or synthesized through organic chemistry. Although traditional production methods (including chemical synthesis and fermentation) enabled the development of a wide range of new chemical entities and many antibiotics, they were not suitable for the production of most proteins. Proteins, or molecules composed of long interlocking chains of amino acids, are simply too large and complex to synthesize feasibly through traditional synthetic chemical methods. Those proteins that were used as therapeutic agents—notably insulin—were extracted from natural sources or produced through traditional fermentation methods. However, since these processes (which were used to produce many antibiotics) could only utilize naturally occurring strains of bacteria, yeast, or fungi, they were not capable of producing the vast majority of proteins. Cohen and Boyer's (publicly funded) key contribution was the invention of a method for manipulating the genetics of a cell so that it could be induced to produce a specific protein. This invention made it possible for the first time to produce a wide range of proteins synthetically and thus opened up an entirely new domain of search for new drugs—the vast store of more than 500,000 proteins that the body uses to carry out a wide range of biological functions.

In principle these new techniques of genetic engineering thus opened up an enormous new arena for research. However the precise function of the majority of these proteins is still not well understood, and the first firms to exploit the new technology chose to focus on proteins such as insulin, human growth hormone, tPA, and Factor VIII —for which scientists had a relatively clear understanding of the biological processes in which they were involved and of their probable therapeutic effect. This knowledge greatly simplified both the process of research for the first biotechnology-based drugs and the process of gaining regulatory approval. It also made it much easier to market the drugs since their effects were well known and a preliminary patient population was already in place.

As firms and researchers gain experience with the new science, however, it has had increasingly dramatic impacts on the ways in which new drugs are discovered. For example the techniques of genetic engineering allow researchers to clone target receptors, so that firms can screen against a pure target rather than against, for example, a solution of pulverized rat brains that probably contain the receptor. They can also allow for the breeding of rats or mice that have been genetically altered to make them particularly sensitive to interference with a particular enzymatic pathway. Both of these techniques allow for the design of greatly improved "screens" against which compounds can be tested for therapeutic activity.

A second strategy has been to focus on a specific disease or condition and to attempt to find a protein that might have therapeutic effects. Here detailed knowledge of the biology of specific diseases is an essential foundation for an effective search. For example researchers working in cancer, AIDS, and autoimmune diseases have focused on trying to discover the proteins responsible for modulating the human immune system. A third strategy is to focus on genomics—the use of knowledge of the human genetic code to uncover new treatments for disease. This strategy is only at the most preliminary stage, but it promises to revolutionize the treatment of many diseases.

Taken together, these events have moved public research from an important but distant foundation for drug discovery to a critically important source of immediately useful knowledge and techniques that is actively engaged by the private sector. Table 1.3 and figures 1.3 and 1.4 graphically illustrate this transition. Table 1.3 summarizes detailed case

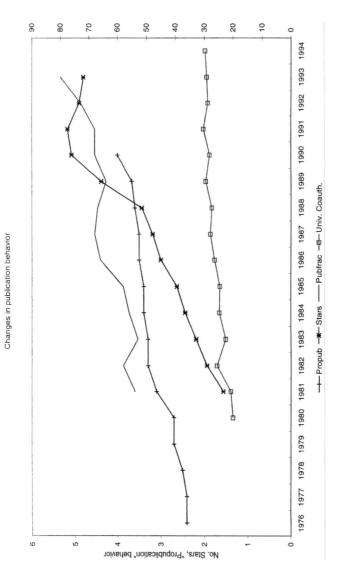

Figure 1.3
The changing relationship between the public and private sector

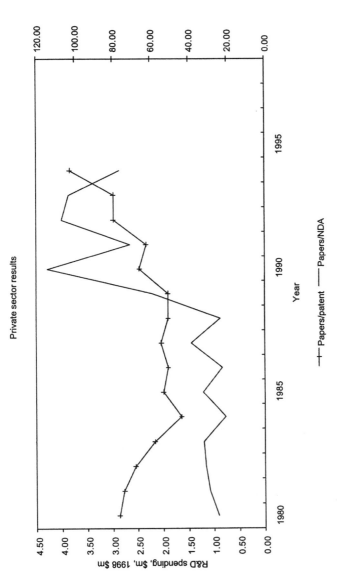

Figure 1.4
The increasing importance of publications

Table 1.3
History of the development of the 21 drugs with highest therapeutic impact introduced between 1965 and 1992

Generic name	Trade name	Indication	Date of key enabling discovery	Public?	Date of synthesis of compound	Public?	Date of market introduction	Lag from enabling discovery to market introduction
Old Fashioned or "random" drug discovery: screening of compounds in whole or partial animal screens								
Cyclosporine	Sandimmune	Immune suppression	NA		1972	N	1983	
Fluconazole	Diflucan	Anti-fungal	1978	N	1982	N	1985?	7
Foscarnet	Foscavir	CMV Infection	1924	Y	1978	Y	1991	67
Gemfibrozil	Lopid	Hyperlipidemia	1962	N	1968	N	1981	19
Ketoconazole	Nizoral	Anti-fungal	1965	N	1977?	N	1981	16
Nifedipine	Procardia	Hypertension	1969	N	1971	N	1981	12
Tamoxifen	Nolvadex	Ovarian cancer	1971	Y	NA		1992	21
"Mechanism driven" research: screening of compounds against a very specific known or suspected mechanism								
AZT	Retrovir	HIV	Contentious	Y	1963	Y	1987	16
Catopril	Capoten	Hypertension	1965	Y	1977	N	1981	16
Cimetidine	Tagamet	Peptic Ulcer	1948	Y	1975	N	1977	29
Finasteride	Proscar	BPH	1974	Y	1986	N	1992	18
Fluoxetine	Prozac	Depression	1957	Y	1970	N	1987	30
Lovastatin	Mevacor	Hyperlipedimia	1959	Y	1980	N	1987	28
Omeprazole	Prilosec	Peptic Ulcers	1978	N	1982		1989	11
Ondansetron	Zofran	Nausea	1957	Y	1983	N	1991	34
Propranolol	Inderol	Hypertension	1948	Y	1964	N	1967	19
Sumatriptan	Imitrex	Migraine	1957	Y	1988	N	1992	35
Drugs discovered through fundamental science								
Acyclovir	Zovirax	Herpes	—		—	Y	1982	
Cisplatin	Platinol	Cancer	1965	Y	1967	Y	1978	13
Erythropoietin	Epogen	Anemia	1950	Y	1985	N	1989	39
Interferon beta	Betaseron	Cancer, others	1950	Y	Various	N	Various	

Authors' compilations.

histories of the discovery and development of 21 drugs identified by two leading industry experts as "having had the most impact upon therapeutic practice" between 1965 and 1992. The table confirms the important role that the public sector plays in providing fundamental insights in basic knowledge as a basis for drug discovery.[5] Only five of these drugs, or 24%, were developed with essentially no input from the public sector. (This contrasts with Maxwell and Eckhardt's finding (Maxwell and Eckhardt 1990) that 38% of their sample of older drugs were developed with no public sector input.) In the second place, these data are consistent with the hypothesis that public sector research has become more important to the private sector over time. The table groups the drugs into three classes according to the research strategy by which they were discovered: those discovered by random screening, those discovered by mechanism-based screening, and those discovered through fundamental scientific advances. Broadly speaking, the degree of reliance on the public sector for the initial insight increases across the three groups, and as the industry has moved to a greater reliance on the second and third approaches, so the role of the public sector has increased. In the first group of therapies—those discovered through "random screening"—public sector researchers made the key enabling discovery in only two of the five drugs. In the two more recent groups public sector researchers made the key discovery in all but two of the cases. The very long lags apparent in the table between fundamental advances in science and their incorporation in marketed products may be shortening as the public and private sectors draw closer together, but it is difficult to draw strong conclusions from this small sample.

One way to capture interaction between the public sector and industry is via the paper trail of publications by pharmaceutical company researchers in the open literature. Publication is a key indicator of participation in the wider scientific community, and in our studies of the management of research in a sample of major pharmaceutical firms, we found evidence from analysis of these "bibliometric" data that this participation has become more and more significant over time. Figure 1.3 shows four key measures of this dimension of the relationship between the public and private sector in the industry, and tracks their evolution over time. *"Propub"* is a measure of the degree to which the firm relies on its scientists' standing in relationship to public science as a key criteria in promotion decisions (Henderson and Cockburn 1994).[6] *"Stars"* is the average number of scientists at each

firm who publish more than 25 papers within any given three year period. *"Pubfrac"* is the percentage of all those scientists whose names appear on a patent in any given year whose name also appears on
scientific publication.[7] *"Univ-coauth"* is the average percentage of the
firm's papers that are coauthored with university authors.[8] All of these
measures increase significantly over the period, illustrating graphically
the private sector's increasing engagement with the world of publicly
available (and largely publicly funded) research. Figure 1.4 illustrates
one result of this dynamic: the number of papers per patent and papers
per NDA (New Drug Application) has also steadily increased over the
period.

V.   What, then, can we say?

The estimation of the effect of publicly funded research on the productivity of the pharmaceutical industry thus presents formidable challenges. It is very difficult to accurately measure either inputs or
outputs: there are very long and highly variable lags in the relationship
between inputs and outputs, and furthermore the nature of this relationship has likely changed dramatically over time.

   Research in this area has thus proceeded along three lines. The first is
the broad brush comparison of the United States with the rest of the
world. The second, perhaps not surprisingly, is the detailed case study.
The third is econometric or statistical. All three suffer from limitations,
but taken together they suggest that publicly funded research has a
very significant impact on the generation of new drugs.

Regional Comparisons

One of the intriguing aspects of the revolution in molecular biology is
that despite the fact that it is global in nature, and despite the fact that
scientific advances are normally thought of as creating a "free good,"
or as being instantaneously available worldwide, it has resulted in
quite different changes in industry structure in different parts of the
world. In the U.S., it has spawned both the emergence of radically new
actors—the new specialized biotechnology firms—and the gradual creation of biotechnology programs within established firms. In Europe,
responses have differed dramatically from country to country. Despite
a strong research tradition in molecular biology, in general Europe has
not witnessed the creation of a specialized biotechnology sector. Several of the leading Swiss and British "Big Pharma" incumbent firms

have attempted to build strong biotechnology capabilities through a combination of internal development and an aggressive program of external acquisition, but the French, German, and Italian firms have been much slower to adopt the new techniques. In Japan, where historically the pharmaceutical industry has been somewhat less innovative than its Western rivals, most substantial investments in biotechnology have been made by firms with historical strengths in fermentation based industries, and the large pharmaceutical companies have been particularly slow to embrace the new technology.

The question of why the phenomenon of the small, independently funded biotechnology startup was initially an American one is an old and much discussed question. One of the reasons that it cannot be answered definitively is that the answer is to a large degree over determined. In the United States a combination of factors made it possible for small, newly founded firms to take advantage of the opportunities created by biotechnology.

On the one hand, the majority of the American biotechnology startups were tightly linked to university departments, and the very strong state of American academic molecular biology clearly played an important role in facilitating the wave of startups that characterized the eighties (Zucker, Darby, and Brewer 1997). The strength of the local science base may also be responsible, within Europe, for the relative British advantage and the relative German and French delay. Similarly the weakness of Japanese industry may partially reflect weakness of Japanese science. There seems to be little question as to the superiority of the American and British scientific systems in the field of molecular biology, and it is tempting to suggest that the strength of the local science base provides an easy explanation for regional differences in the speed with which molecular biology was exploited as a tool for the production of large molecular weight drugs.

On the other hand, a number of other important factors supported the new firms' growth. These factors included a favorable financial climate, strong intellectual property protection, a scientific and medical establishment that could supplement the necessarily limited competencies of small newly founded firms, a regulatory climate that did not restrict genetic experimentation, and, perhaps most importantly, a combination of a very strong local scientific base with academic and cultural norms that permitted the rapid translation of academic results into competitive enterprises. Nelson (1993) has labeled this a "national system of innovation," and it appears to have been particularly conducive to innovation in biotechnology. In Europe (although to a lesser extent in the U.K.) and in Japan many of these factors were not in place. For example, for many years the patentability of various aspects of biotechnology was uncertain in Europe, and until recently there was a relatively small local venture capital industry. In general, it was left to larger firms to exploit the new technology in these countries.

Case Study Research   There have also been a significant number of careful case studies of this issue, most focused on the development of detailed histories of the discovery of new drugs. See for example Borel, Kis, and Beveridge 1995, Comroe and Dripps 1976, Penan 1996, Raiten and Berman 1993, Richardson et al. 1990, and Rittmaster 1994. By tracing the involvement of particular individuals or laboratories in the discovery of a particular drug it is possible, at least in principle, to identify and evaluate the relative importance of privately funded versus publicly funded research. Of course, the exercise can be very difficult in practice, and is unlikely to produce unambiguous conclusions.

Consider the case of AZT, the first drug to approved by the FDA for use in treatment of HIV infection. AZT was first synthesized in the early 1960s by a public sector researcher looking for activity against cancer. It then languished for many years in the library of compounds maintained by antiviral researchers at Burroughs Wellcome. Its value in prolonging the life of some AIDS patients only became apparent when BW sent it, along with a dozen other candidate compounds, for testing against a screen developed at NIH. BW then took the lead in conducting clinical trials and obtaining FDA approval. "Who discovered what and when" was an integral part of the intense controversy surrounding this case, with the U.S. Supreme Court eventually ruling against claims that NIH scientists should have been listed as inventors on BW's patents on the use of AZT in treatment of AIDS.

Legal claims aside, debates about priority in discovery are an integral part of science, and different observers may place more or less weight on different contributions. In many instances, it is simply impossible to definitively assign credit for the invention of a drug to a specific individual or institution. Furthermore, many of these case histories overlook the subtler influences of the public sector in providing "infrastructure," graduate training, and so forth. But between them these studies—and others like them—make a compelling quantitative case for the importance of publicly funded research. All of them suggest that publicly funded research made critical contributions to the discovery of an important therapeutic advance.

Econometric or Statistical Studies   Econometric studies of the impact of publicly funded research on private sector productivity supplement the particularity of case studies with more general results, but are subject to all of the problems that we outlined above. Figures 1.5 and 1.6 hint at some of the issues that must be dealt with in interpreting the raw quantitative data. Both figures show that several key measures of the output of the industry—papers, patents, and NDAs, or New Drug Approvals—have been increasing over time.[9] But all three measures involve considerable error,[10] all three measures trend up quite smoothly over time, and all three are only loosely related to social impact. Presumably we care about patient health, not papers, patents, or NDAs *per se*, and while there is almost certainly *some* link between the two, at any general level it is impossible to be precise about what it might be. Similarly there are the long lags to consider: the NDAs approved tomorrow will rest on research that was performed anywhere from 5 to 15 years ago, and since both private and public research trends up over time it is very difficult to separately identify their effects.

Despite these difficulties, a number of researchers have attempted to measure the effects of public sector research directly. Zucker, Darby, and Brewer (1997) show that biotechnology startups tend to co-locate with public sector researchers, an intriguing and suggestive result. Zucker, Darby, and Armstrong (1998) show further that collaborations between these new firms and university stars is correlated with some measures of success. For an average firm, five articles coauthored by academic stars and the firm's scientists imply about five more products in development, 3.5 more products on the market, and 860 more employees. These results are consistent with the hypothesis that university research has a powerful effect on the private sector, though they should be interpreted carefully—the authors were not able to fully control for the level of R&D spending by the firms or the quality of their other scientists.

Two studies have explored another indirect measure of public sector impact: the relationship between a firm's ability to take advantage of knowledge generated in the public sector and its own productivity. Gambardella (1995) showed that in the 1970s and 1980s those pharmaceutical firms that published more scientific papers were relatively more productive than their rivals. In a similar vein, in Cockburn and Henderson (1998) we explored the relationship between a firm's research productivity and its "connectedness" to the public sector, using data on coauthorship of scientific papers across institutional

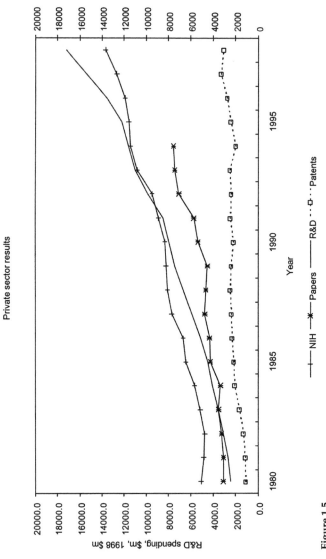

Figure 1.5
Patents and papers as the output of pharmaceutical research

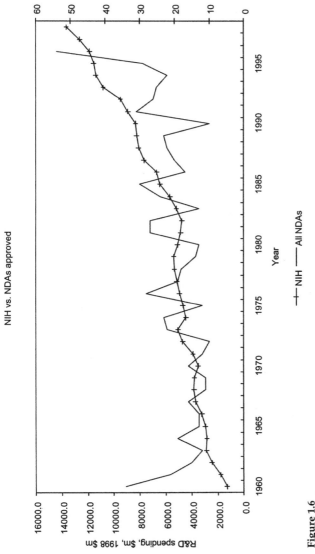

Figure 1.6
NDA output over time

boundaries. "Connectedness" in this sense is closely related to a number of other factors that also increase the productivity of privately funded pharmaceutical research, including the number of star scientists employed by the firm and the degree to which the firm uses standing in the public rank hierarchy as a criterion for promotion. Linking these data with measures of research productivity we found that "connectedness" and research performance are correlated across firms and over time. While any estimate of this type must be treated with great caution, our results also suggested that differences in the effectiveness with which a firm was accessing the upstream pool of knowledge corresponded to differences in the research productivity of firms in our sample of as much as 30%.

One interpretation of this result is that it represents a lower bound estimate of the impact of public sector research, since by definition it excludes the impact of any publicly generated knowledge that can be costlessly accessed across the industry. However the fact that "connectedness" is likely to be correlated with other hard-to-observe organizational practices that improve research productivity, as well as other important sources of unobserved heterogeneity across firms (such as the quality of human capital) made us hesitate to assign the result too much weight. Rather we suggested that our results were consistent with the hypothesis that the ability to take advantage of knowledge generated in the public sector requires investment in a complex set of activities that taken together change the nature of private sector research. They thus raise the possibility that the *ways* in which public research is conducted may be as important as the *level* of public funding. To the extent that efforts to realize a direct return on public investments in research lead to a weakening of the culture and incentives of "open science," our results are consistent with the hypothesis that the productivity of the whole system of biomedical research may suffer.

In a study at a more aggregate level, Ward and Dranove (1995) showed that a 1% increase in research funding by the National Institutes of Health leads to an estimated 0.6–0.7% increase in spending by members of the Pharmaceutical Manufacturers of America (PMA), after a lag of 6 to 10 years. This result is also consistent with the hypothesis that the private return to publicly funded research is quite high, since if increases in public sector research fuel private sector increases, then presumably the presence of public sector research is raising the marginal productivity of private sector work.

The most recent paper in this stream of research is by Andrew Toole (1999). Toole uses data at the level of the therapeutic class to obtain estimates of the rate of return to publicly funded research. His (unpublished) estimates imply that a 1% increase in the stock of public basic research ultimately leads to a 2.0% to a 2.4% increase in the number of commercially available new compounds, and that industry firms appropriate a return on public science investment in the range of 11% to 32%. He notes that this result suggests that the returns to public science are actually rather larger since these estimates are based on conservative estimates of firm profits from an average compound and since they ignore any consumer surplus that may be created by the introduction of a new therapy.

Conclusions   We have suggested that there are a number of factors that make it difficult to estimate precisely the impact of publicly funded research. Such estimation is always difficult, but in the case of the pharmaceutical industry it is a notably difficult task since the public sector provides inputs to industry research in so many different and subtle ways, and spillovers are likely to be so large that the social returns to innovation are substantially different from the measurable private returns.

Nevertheless, a considerable body of both qualitative and quantitative evidence indicates that the public sector has had a profoundly positive impact on the industry, and that this appears to have increased significantly over time. Qualitative evidence suggests that public sector research has made possible fundamental advances in the ways in which new drugs are discovered and has opened up doors that may revolutionize the treatment of disease. The quantitative evidence suggests that the rate of return to public sector research *as measured by its effect on the private sector,* may be as high as 30%.

There are a number of reasons for believing that this figure is in fact a quite conservative estimate of the overall social return to publicly funded research in this sector of the economy. First, it is highly unlikely that private sector firms capture all of the benefits to public sector research in their own output. When drugs come off patent, to take one example, their price tends to fall considerably, but the benefit to consumers in terms of QUALYs or other measures of health status remains constant. Thus "true" output is likely to be seriously undercounted, and economic estimates of the bang for the buck from publicly funded research will therefore, if anything, be lower bounds to the real value.

Second, the lag between when basic scientific advances are made and when their impact becomes visible in marketed products is particularly long in this industry. Despite intense commercial competition and the dedicated effort of many thousands of individuals, it can take 10 years or more for promising discoveries to be turned into approved drugs. Today's improvements in treatment of many diseases reflect public research expenditures made in the 1960s and 1970s. The bulk of the impact of the investments made in the 1980s and 1990s has probably not yet been felt. Arguably, we have yet to benefit from the most important contribution of modern publicly funded science: the breakthrough in our understanding of genetics and molecular biology that is summarized under the name "the biotechnology revolution." There are hundreds of new compounds in development that draw upon this knowledge, and surely thousands more yet to be discovered. Econometric studies conducted 10 or 20 years from now are likely to find even higher rates of return to publicly funded research.

Over the past 50 years, the pharmaceutical industry and the publicly funded biomedical research establishment have grown hand in hand in their size and economic significance. Ever larger investments in research on both sides have resulted in new drugs and vaccines that are responsible for very significant improvements in health and well-being. This remarkable innovative performance is unlikely to have been realized without substantial public support of basic research, along with the development of close linkages between private sector and public sector institutions.

The relationships between the NIH, government labs, universities, and the private sector continue to evolve, and areas of conflict have inevitably arisen. In genome research, for example, private firms have been seeking proprietary rights over some of the results of decades of publicly funded work on DNA sequencing. Equally, universities have become increasingly aggressive and effective in realizing licensing revenue from their discoveries. These changes are altering the delicate balance between nonprofit and for-profit institutions which appears to have been so effective in the past at generating scientific advances and bringing them to market, and are surely a cause for some concern.

Nonetheless, absent any evidence of exhaustion of scientific opportunities, there is a compelling case for continued substantial public support of the biomedical sciences. As today's taxpayers reach retirement age they will enjoy a generous return from these investments, but if the experience of the past five decades is any guide it will be their children and their children's children who will benefit the most.

Notes

This paper was prepared for the NBER Conference on Science and Public Policy, Washington, D.C., April 2000. This study was funded by POPI, the Program for the Study of the Pharmaceutical Industry at MIT and by the MIT Center for Innovation in Product Development under NSF Cooperative Agreement Number EEC-9529140. This support is gratefully acknowledged. Jeff Furman provided outstanding research assistance. cockburn@bu.edu, rhenders@mit.edu.

1. The preferred embodiment of this idea expresses productivity in terms of a production function which models output as a function of inputs. These functions can be quite elaborate, allowing for returns to scale, substitution between inputs, etc. Estimation of these functions raises a further set of problems, see, e.g., Griliches, 1979, 1994, 1995.

2. The vagueness of this statement reflects the very considerable methodological problems inherent in these types of studies. To give a taste of these, consider the statistical problems inherent in trying to econometrically estimate production functions in which measures of publicly funded knowledge capital appear as an input. Even if accurate measures of inputs and outputs can be found, getting accurate estimates of the parameter values which tell us about the rate of return to public research is very difficult. For example, in general, measures of outputs and inputs tend to be correlated with each other and to move together over time. This makes it hard to determine the direction of causality: does research cause sales or do sales cause research? Even if there is sufficient independent variation in these variables, it is far from clear what the appropriate functional form of the production function might be.

3. Most major pharmaceuticals are multinational, performing R&D in more than one country. These figures do not include an additional 10–20% of overseas R&D spending by U.S.-headquartered firms, or expenditures in the U.S. by foreign-based firms.

4. Indeed the development of "combinatorial chemistry" coupled with the techniques of "high throughput screening" have given a new lease of life to random drug discovery.

5. For purposes of general comparison we list a date of key enabling discovery for each drug. The choice of any particular event as the key enabling discovery is bound to be contentious, since in pharmaceuticals, as in many fields, discovery usually rests on a complex chain of interrelated events. In the case of drugs discovered through screening we give the date of first indication of activity in a screen. In the case of mechanism based drugs, we give the date of the first clear description of the mechanism. Dates for the third class are only broadly indicative, and all should be used carefully.

6. "Propub" was constructed using detailed qualitative data at 10 major pharmaceutical firms. For details, see Henderson and Cockburn 1994.

7. "Stars" and "Pubfrac" were constructed using publicly available data from 19 large pharmaceutical firms. For details, see Cockburn, Henderson, and Stern 2000.

8. This variable is constructed for the same sample of 10 major firms for which "Propub" was constructed, using publicly available data. See Cockburn and Henderson 1998.

9. Industry sales have also been increasing, at roughly the same rate as private R&D spending. Recall, however, that the lag between R&D spending and the generation of sales is a very long one!

10. We show here only NDAs that warrant Class 1 or Class 2 ranking by the FDA—entirely new therapies of considerable merit and therapies that are essentially equivalent to existing therapies.

References

Adams, J. 1990. "Fundamental Stocks of Knowledge and Productivity Growth." *Journal of Political Economy* 98(4): 673–702.

Agarwal, A. 2000. "The Diffusion of Knowledge from University to Industry: A Study at MIT." Forthcoming Ph.D. Dissertation, University of British Columbia.

Arrow, K. 1962. "Economic Welfare and the Allocation of Resources for Invention." In R. Nelson, ed., *The Rate and Direction of Inventive Activity.* Princeton, NJ: Princeton University Press: 609–19.

Berndt, E. R., D. Cutler, R. G. Frank, Z. Griliches, J. P. Newhouse and J. E. Triplett. 2000. "Price Indexes for Medical Care Goods and Services: An Overview of Measurement Issues." Forthcoming in David Cutler and Ernst R. Berndt, Eds., *Medical Care Output and Productivity,* Chicago: University of Chicago Press for the National Bureau of Economic Research.

Borel, J. F., Z. L. Kis, and T. Beveridge. 1995. "The History of the Discovery and Development of Cyclosporine." In V. J. Merluzzi and J. Adams, eds., *The Search for Anti-Inflammatory Drugs.* Boston: Birkhauser.

Clymer, H. A. 1975. "The Economic and Regulatory Climate: U.S. and Overseas Trends." In R. B. Helms, ed., *Drug Development and Marketing.* Washington, DC: American Enterprise Institute.

Cockburn, I., and R. Henderson. 1998, June. "Absorptive Capacity, Coauthoring Behavior, and the Organization of Research in Drug Discovery." *Journal of Industrial Economics* XLVI(2): 157–82.

Cockburn, I. and A. Anis. 2000. "Hedonic Analysis of Arthritis Drugs." Forthcoming in David Cutler and Ernst R. Berndt, Eds., *Medical Care Output and Productivity,* Chicago: University of Chicago Press for the National Bureau of Economic Research.

Cockburn, I., R. Henderson, and S. Stern. 1999, January. "Balancing Incentives: The Tension between Basic and Applied Research." Working Paper no. 6882, National Bureau of Economic Research, Cambridge, MA.

Cockburn, I., Henderson, R. and S. Stern. Fall 2000. "Untangling the Origins of Competitive Advantage," *Strategic Management Journal,* 21: 1123–1145.

Comroe, J., and R. Dripps. 1976. "Scientific Basis for the Support of Biomedical Research." *Science* 192: 105.

Cutler, D., and E. Richardson. 1999. "Your Money and Your Life: The Value of Health and What Affects It." A. Garber, ed. *Frontiers in Health Policy Research.* National Bureau of Economic Research, vol. 2, Cambridge, MA: MIT Press.

Dasgupta, P., and P. A. David. 1987. "Information Disclosure and the Economics of Science and Technology." In G. R. Feiwel, ed., *Arrow and the Ascent of Modern Economic Theory*. New York: NYU Press.

Dasgupta, P., and P. A. David. 1994. "Towards a New Economics of Science." *Research Policy* 23: 487–521.

David, P. A., D. Mowery, and E. Steinmuller. 1992. "Analyzing the Economic Payoffs from Basic Research." *Economics of Innovation and New Technologies* 2(4): 73–90.

Galambos, L., and J. E. Sewell. 1996. *Networks of Innovation: Vaccine Development at Merck, Sharp and Dohme, and Mulford, 1895–1995.* Cambridge: Cambridge University Press.

Gambardella, A. 1995. *Science and Innovation: The US Pharmaceutical Industry in the 1980s.* Cambridge, U.K.: Cambridge University Press.

Griliches, Z. 1979. "Issues in Assessing the Contribution of Research and Development to Productivity Growth." *Bell Journal of Economics* 10(1): 92–116.

Griliches, Z. 1994. "Productivity, R&D and the Data Constraint." *American Economic Review* 84(1): 1–23.

Griliches, Z. 1995. "R&D and Productivity: Econometric Results and Measurement Issues." In P. Stoneman, ed., *Handbook of the Economics of Innovation and Technical Change.* Oxford, U.K.: Blackwell: ch 3, pp 52–71.

Henderson, R., and I. M. Cockburn. 1994. "Measuring Competence? Exploring Firm Effects in Pharmaceutical Research." *Strategic Management Journal* 15: 63–84.

Henderson, R., and I. M. Cockburn. 1996. "Scale, Scope and Spillovers: The Determinants of Research Productivity in Drug Discovery." *RAND Journal of Economics* 27(1): 32–59.

Irvine, J., B. Martin, and P. Isard. 1990. *Investing in the Future: An International Comparison of Government Funding of Academic and Related Research.* Aldershot, England: Edward Elgar Publishers.

Jones, C., and J. Williams. 1995. "Too Much of a Good Thing? The Economics of Investment in R&D." Working Paper no. 96-005, Stanford University Department of Economics. Stanford, CA.

Laragh, J. H., et al. 1972. "Renin, Angiotensin and Aldosterone System in Pathogenesis and Management of Hypertensive Vascular Disease." *American Journal of Medicine* 52: 644–52.

Mansfield, E. 1991. "Academic Research and Industrial Innovation." *Research Policy* 20(1): 1–12.

Maxwell, R. A., and S. B. Eckhardt. 1990. *Drug Discovery: A Case Book and Analysis.* Clifton, NJ: Humana Press.

Merton, D. 1973. "On the Sociology of Science." In N. W. Starer, ed., *The Sociology of Science: Theoretical and Empirical Investigation,* Chicago: University of Chicago Press.

National Science Foundation. 1968, December. "Technology in Retrospect and Critical Events in Science" Unpublished manuscript prepared by IIT Research Institute. NSF C535.

Nelson, R. R. 1959. "The Simple Economics of Basic Scientific Research." *Journal of Political Economy* 67(2): 297–306.

Nelson, R. R. ed. 1993. *National Innovation Systems: A Comparative Analysis*. Oxford, U.K.: Oxford University Press.

Penan, H. 1996. "R&D Strategy in a Techno-Economic Network: Alzheimer's Disease Therapeutic Strategies." *Research Policy* 25: 337–58.

Raiten, D., and S. Berman. 1993. "Can the Impact of Basic Biomedical Research be Measured? A Case Study Approach." Working Paper, Life Sciences Research Office, Federation of American Societies for Experimental Biology. Washington, D.C.

Richardson, K., K. Cooper, M. S. Marriott, M. H. Tarbit, P. F. Troke, and P. J. Whittle. 1990. "Discovery of Fluconazole, a Novel Antifungal Agent." *Review of Infectious Diseases* 12(3): S267–71.

Rittmaster, R. 1994, January 13. "Finasteride." *New England Journal of Medicine* 120–5.

Stephan, P. 1996, September. "The Economics of Science." *Journal of Economic Literature* 34: 1199–235.

Stern, S. 1999, October. "Do Scientists Pay To Be Scientists?" Working Paper no. 7410, National Bureau of Economic Research, Cambridge, MA.

Toole, A. 1999, November. "The Contribution of Public Science to Industrial Innovation: An Application to the Pharmaceutical Industry." Discussion Paper no. 98-6, Stanford Institute for Economic Policy Research. Stanford, CA.

Ward, M., and D. Dranove. 1995. "The Vertical Chain of R&D in the Pharmaceutical Industry." *Economic Inquiry* 33: 1–18.

Wurtman, R. J., and R. L. Bettiker. 1994. "How to Find a Treatment for Alzheimer's Disease." *Neurobiology of Aging* 15: S1–3.

Wurtman, R. J., and R. L. Bettiker. 1996, Spring. "Training the Students Who Will Discover Treatments for Psychiatric Diseases." *Psychiatric Research Report*.

Zucker, L., M. Darby, and M. Brewer. 1997. "Intellectual Human Capital and the Birth of U.S. Biotechnology Enterprises." *American Economic Review* 88(1): 290–306.

Zucker, L., M. Darby, and J. Armstrong. 1998. "Geographically Localized Knowledge: Spillovers or Markets?" *Economic Inquiry* 36: 65–86.

2

Creating Markets for New Vaccines
Part I: Rationale

Michael Kremer, *Harvard University, The Brookings Institution, and NBER*

Executive Summary

Malaria, tuberculosis, and the strains of HIV common in Africa kill approximately five million people each year. Yet research on vaccines for these diseases remains minimal—largely because potential vaccine developers fear that they would not be able to sell enough vaccine at a sufficient price to recoup their research expenditures.

Enhancing markets for new vaccines could create incentives for vaccine research and increase accessibility of any vaccines developed. For example, the President of the World Bank has proposed establishing a fund to help developing countries finance purchases of specified vaccines if they are invented. The U.S. administration's 2000 budget proposal includes a tax credit for new vaccines that would match each dollar of vaccine sales with a dollar of tax credits. This paper examines the rationale for such proposals.

Private firms currently conduct little research on vaccines against malaria, tuberculosis, and the strains of HIV common in Africa. This is not only because these diseases primarily affect poor countries, but also because vaccines are subject to severe market failures. Once vaccine developers have invested in developing vaccines, governments are tempted to use their powers as regulators, major purchasers, and arbiters of intellectual property rights to force prices to levels that do not cover research costs. Research on vaccines is an international public good, and none of the many small countries that would benefit from a malaria, tuberculosis, or HIV vaccine has an incentive to encourage research by unilaterally offering to pay higher prices. In fact, most vaccines sold in developing countries are priced at pennies per dose, a tiny fraction of their social value. More expensive, on-patent vaccines are typically not purchased by the poorest countries. Hence, private developers lack incentives to pursue socially valuable research opportunities. Large public purchases could potentially enlarge the market for vaccines, benefiting both vaccine producers and the public at large.

Government-directed research programs may be well suited for basic research, but for the later, more applied stages of research, committing to compensate successful private vaccine developers has important advantages. Under such programs, the public pays only if a successful vaccine is actually developed. This gives pharmaceutical firms and scientists strong incentives to self-select research projects that have a reasonable chance of leading to a vaccine, and to focus on developing a viable vaccine rather than pursuing other goals.

Committing to purchase vaccines and make them available to poor countries may be attractive relative to other ways of rewarding vaccine developers. Extending patents on other pharmaceuticals to reward developers of new vaccines would place the entire burden of financing vaccines on those needing these other pharmaceuticals. Increasing prices for current vaccines without explicit incentives for development of new vaccines would be insufficient to spur new research.

I. Introduction

Malaria, tuberculosis, and the strains of HIV prevalent in Africa kill almost five million people each year. Yet relative to this enormous burden, very little vaccine research is directed toward these diseases. Potential vaccine developers fear that they would not be able to sell enough vaccine at a high enough price to recoup their research investments. This is both because these diseases primarily affect poor countries, and because vaccine markets are severely distorted. This paper examines the economic rationale for committing in advance to purchase vaccines for these diseases. Such commitments could create incentives for vaccine research and help ensure that if vaccines were developed, poor countries could afford them. Because a vaccine purchase commitment would require no funds until a vaccine was available, it would not compete with budgets for current efforts to control diseases using existing technology.

These issues are particularly timely. The U.S. administration's budget proposal (available at http://www.treas.gov/taxpolicy/library/grnbk00.pdf) includes $1 billion in tax credits over the 2002–2010 period for vaccine sales. The program would match every dollar of qualifying vaccine sales with a dollar of tax credit, effectively doubling the incentive to develop vaccines for neglected diseases. Qualifying vaccines would have to cover infectious diseases which kill at least one million people each year, would have to be approved by the U.S. Food and Drug Administration (FDA), and would have to be certified by the Secretary of the Treasury after advice from the U.S. Agency for International Development (USAID). To qualify for the tax credit, sales would have to be made to approved purchasing institutions, such as the United Nations Children's Fund (UNICEF). Although the administration's proposal is structured as a tax credit, it would have effects similar to an expenditure program that matched private funds spent on vaccines.

The World Bank president, James Wolfensohn, recently said that the institution plans to create a $1 billion fund to help countries purchase specified vaccines if and when they are developed (Financial Times, 2000). Wolfensohn's proposal is being discussed within the Bank and would have to be approved by the Bank's board. One option under consideration is a more general program to combat communicable diseases of the poor. For a general program to stimulate research, it must include an explicit commitment to help finance the purchase of new vaccines if and when they are developed. Without an explicit commitment along the lines proposed by Wolfensohn, it is unlikely that the large scale investments needed to develop vaccines will be undertaken.

The concept of a vaccine purchase fund has also received support from European political leaders (http://www.auswartiges.amt.de 1999, DFID 2000).

Section II of this paper provides background information on malaria, HIV, and tuberculosis; discusses the prospects for vaccines for these diseases; and reviews the current state of scientific progress toward vaccine development.

Section III discusses distortions in the market for vaccines and for vaccine research. People tend to underconsume vaccines for a number of reasons. First, individuals have inadequate incentives to take vaccines, since those who take vaccines not only benefit themselves, but also benefit others by breaking the cycle of infection. Second, the chief beneficiaries of vaccination are often children, who cannot contract to pay vaccine sellers the future earnings they will reap if they take vaccines and stay healthy. Third, consumers are often more willing to pay for treatment than prevention, perhaps because it takes time for them to learn about the effectiveness of vaccines. Monopoly pricing further limits access to patented vaccines. Perhaps because of these factors, most countries purchase vaccines in bulk and distribute them at subsidized rates. At appropriate prices, these large public purchases could potentially make both vaccine producers and the population at large better off than they would be under monopoly pricing by reducing the cost per dose and expanding the market.

Distortions in the market for vaccine research are even greater than those for vaccines themselves. Rough calculations suggest that the social benefits of malaria, tuberculosis, or HIV vaccines may easily exceed the returns to a private developer by a factor of 10 or more, so vaccine developers will lack incentives to pursue socially valuable research opportunities. Research incentives are too small in many fields, but the situation is particularly problematic for vaccines and is dire for vaccines against diseases that primarily affect poor countries. It is often possible to design around vaccine patents, and since vaccines are primarily sold to governments, brand loyalty provides minimal benefit to the original developer. Once developers have sunk resources into developing vaccines, governments are often tempted to use their powers as regulators, major purchasers, and arbiters of intellectual property rights to obtain vaccines at prices which cover only manufacturing costs, not research costs. Since research and development on vaccines for malaria, tuberculosis, and HIV is a global public good that benefits many small countries, no single country has an incentive to encourage research by offering higher prices, and hence many countries have historically provided little or no intellectual property rights protection to vaccines. Most vaccines sold in developing countries sell for pennies per dose, and newer, on-patent, vaccines, which sell for a dollar or two per dose, do not reach the poorest countries. Crude calculations suggest that a malaria vaccine would be cost-effective relative to other developing country health programs at $41 per person immunized. The gap between the $41 at which a vaccine would be cost-effective and the $2 which the historical record suggests a vaccine developer would be lucky to obtain for a vaccine implies that under current institutions, potential vaccine developers would not have incentives to pursue socially valuable research opportunities.

Section IV examines the appropriate roles of "push" and "pull" programs in encouraging vaccine research and improving access to vaccines once they are developed. Push programs pay for research inputs, for example through grants to researchers, while pull programs pay for an actual vaccine. Push programs are well suited to financing basic research, because it is important that the results of basic research are quickly communicated to other scientists. Grant-funded researchers have incentives to publish quickly, while researchers with strong financial incentives to develop a vaccine might wish to withhold information from competitors. Historically, however, governments have relied heavily on push programs to encourage even the later, more applied stages of vaccine development, in part because it was thought necessary to finance research expenditures in advance of the development of a vaccine. With the development of the biotech industry and the increased availability of finance from venture capitalists and large pharmaceutical firms, it is now much easier for scientists to attract investors to finance research, as long as a substantial market is expected for the product.

Pull programs can provide such a market, and they have several attractive features relative to traditional push programs for encouraging the later stages of vaccine development. Under pull programs, the public pays nothing unless a viable vaccine is developed. This gives researchers incentives to self-select projects with a reasonable chance of yielding a viable vaccine, rather than to oversell their research prospects to research administrators and the public. It allows politicians and the public to be confident that they are paying for an actual vaccine, rather than supporting a vaccine-development effort that might not be warranted scientifically. Pull programs also provide strong financial incentives for researchers to focus on developing a marketable vaccine, rather than pursuing other goals, such as publishing academic articles. Finally, pull programs can help ensure that if vaccines are developed, they will reach those who need them.

Section V compares a vaccine purchase commitment program to other pull programs designed to increase incentives for vaccine research. Rewarding vaccine developers with extensions of patents on other pharmaceuticals would inefficiently and inequitably place the entire burden of financing vaccine development on patients who need these other pharmaceuticals. Cash prizes for research are economically similar to a vaccine purchase program, but provide a somewhat weaker link between vaccine quality and the compensation paid to vaccine developers. They are also likely to be politically less attractive and therefore less credible to potential vaccine developers. Encouraging vaccine development through research tournaments is likely to be difficult, since there is no guarantee that a vaccine could be developed within a fixed time period. While expanded purchases and deliveries of currently underutilized vaccines would be highly cost-effective health interventions in their own right, such purchases are unlikely on their own to convince potential developers of vaccines for malaria, tuberculosis, or clades of HIV common in Africa that historically fickle international aid donors will provide funds to purchase vaccines for these diseases 10 or 15 years from now. Explicit purchase commitments would also be needed.

A companion paper, "Creating Markets for New Vaccines: Part II: Design Issues," discusses how commitments to purchase vaccines could be structured.

This paper builds on previous literature. The idea of committing to purchase vaccines was discussed in WHO 1996 and was advocated by a coalition of organizations coordinated by the International AIDS Vaccine Initiative at the 1997 Denver G8 summit. Since then, the idea has been explored by the World Bank AIDS Vaccine Task Force (World Bank 1999, 2000). Kremer and Sachs (1999) and Sachs (1999) have advocated the establishment of a program in the popular press. This paper also draws on earlier work on vaccines, including Batson 1998, Dupuy and Freidel 1990, Mercer Management Consulting 1998, and Milstien and Batson 1994, and on the broader academic literature on research incentives, including Guell and Fischbaum 1995, Johnston and Zeckhauser 1991, Lanjouw and Cockburn 1999, Lichtmann 1997, Russell 1998, Scotchmer 1999, Shavell and van Ypserle 1998, and Wright 1983.

This paper differs from some of the earlier work mentioned in examining the case for commitments to purchase vaccines in light of the underlying economic principles of asymmetric information and time consistency. In particular, this paper argues that information asymmetries between funders and researchers may hamper programs that fund researchers in advance. The time-inconsistent preferences of governments imply that in the absence of specific commitments general statements of intent to purchase vaccines will not be credible. This paper also differs from earlier work in comparing commitments to purchase vaccines to other pull programs.

II.   Background on Malaria, HIV, and Tuberculosis

This section reviews the burden of the major infectious diseases, discusses scientific prospects for vaccines, and argues that current research efforts are paltry relative to the burden these diseases impose.

The Burden of Malaria, HIV/AIDS, and Tuberculosis

Estimates of the burden of infectious disease vary widely, but it is clear that the burden is huge. The World Health Organization estimates that each year there are 300 million clinical cases of malaria and 1.1 million deaths from malaria. Almost all cases are in developing countries, and almost 90% are in Africa (WHO 1999a). Malaria is particularly likely to kill children and pregnant women. Resistance is spreading to the major drugs used for treating malaria and for providing short-term protection to travelers (Cowman 1995).

Each year, approximately 1.9 million people die from tuberculosis. More than 98% of these deaths occur in developing countries (WHO 1999a). However, with up to 17% of tuberculosis infections resistant to all five major anti-tubercular drugs, the spread of resistance poses a threat to developed as well as developing countries (WHO, 1997b). The existing BCG vaccine, which is distributed widely, provides short-term, imperfect protection against tuberculosis, but a more effective vaccine, providing longer-term protection, is lacking.[1]

More than 33 million people are infected with HIV worldwide, over 95% of whom live in developing countries. In 1998, about 2.3 million people died of AIDS, 80% of whom lived in sub-Saharan Africa. Approximately 5.8 million people were newly infected, 70% of whom were in sub-Saharan Africa (WHO 1999a; UNAIDS 1998). New life-extending HIV treatments are far too expensive for most individuals and governments in low-income countries. Since people with compromised immune systems are especially vulnerable to tuberculosis, the spread of HIV is contributing to the spread of tuberculosis. Indeed, of the 1.9 million people who die annually from tuberculosis, 400,000 are infected with HIV.

The Potential for Vaccines

Vaccines have proven effective against many other infectious diseases, and in the long run, they are likely to be the most effective and sustainable way to fight malaria, tuberculosis, and HIV/AIDS. The potential of vaccines is illustrated most vividly by the success of the smallpox vaccination program, which led to the eradication of the disease in the 1970s. About three-quarters of the world's children receive a standard package of cheap, off-patent vaccines through WHO's Expanded Program on Immunization (EPI), and these vaccines are estimated to save 3 million lives per year (Kim-Farley, 1992).[2] However, only a small fraction of children in poor countries receive newer vaccines, such as the *Haemophilus influenzae* type b (Hib) vaccine, which are still on patent and hence more expensive.

The Global Alliance for Vaccines and Immunization (GAVI), with major financing from the Gates Foundation, is undertaking a large-scale effort to improve utilization of existing vaccines. This effort is likely to raise coverage rates and save millions of lives. Coverage rates would likely be further increased if effective vaccines were available against malaria, tuberculosis, or HIV/AIDS, since governments would then have greater incentives to maintain their immunization infrastructure, and parents would have more incentive to bring their children in for vaccination. Even if malaria, tuberculosis, or HIV vaccines only achieved the same coverage rates as the inexpensive EPI vaccines, they would still save millions of lives.

The question of whether vaccines can be developed against malaria, tuberculosis, and HIV remains open, but there is reason to be optimistic. A recent National Academy of Sciences report (1996) concludes that the development of a malaria vaccine is scientifically feasible. Candidate vaccines have been shown to protect against malaria in several rodent and primate models. Moreover, the human immune system can be primed against natural malaria infection. People who survive beyond childhood in malaria endemic areas obtain limited immunity which protects them against severe malaria, although not against parasitemia and milder illness. Since vaccines prime the immune system by mimicking natural infection, vaccines may similarly provide protection against severe disease. Recently, candidate vaccines have been shown to induce protection against tuberculosis infection in animal models. The example of the existing BCG vaccine suggests that the human immune system can be primed against tuberculosis infection. A number of candidate HIV vaccines protect monkeys against infection and induce immune responses in humans.

Nonetheless, formidable scientific and technological obstacles remain in the way of the development of malaria, tuberculosis, and HIV vaccines. All three diseases have many variants and evolve rapidly, making it difficult to design vaccines which are effective against all variants of the disease and which remain effective over time.

Recent advances in immunology, biochemistry, and cloning have given scientists new tools to understand the immune response to these diseases, find correlates of protection useful in testing whether candidate vaccines are likely to succeed, and develop better animal models. Genetic sequencing of the organisms causing tuberculosis, AIDS, and malaria is either complete or far advanced. This may help scientists create vaccines that target many different antigens, and thus are more effective in the face of genetic diversity.

Current Vaccine Research

Despite the increasing scientific potential, current research on vaccines for malaria, tuberculosis, and HIV is paltry relative to the burden of these diseases. According to a Wellcome Trust study, public and nonprofit malaria research amounted to about $84 million in 1993 (Wellcome Trust 1996) with vaccine research making up only a small fraction of the total. The amount of private sector spending on malaria is unknown, but is generally considered to be far lower than public spending. Less is known about total expenditures on tuberculosis research, but the United States National Institutes of Health, one of the world's leading funders of basic research, spends around $65 million per year on tuberculosis research, compared with $2.7 billion on cancer research (NIH 1999).

Applied AIDS research is overwhelmingly oriented toward treatments which would be appropriate for people with AIDS in rich countries, rather than toward vaccines appropriate for poorer countries. The multi-drug treatments for HIV are not feasible for poor countries, since they cost $10,000–16,000 a year (PhRMA 1999), require ongoing immune monitoring, and need to be taken in perpetuity according to a precise protocol. To the extent that vaccine research is conducted, it is primarily oriented toward the HIV strains common in rich countries. Most candidate HIV vaccines tested worldwide are based on clade E, the strain of the virus most widespread in the United States, Europe, Australia, and Latin America, rather than the clades most common in Africa, where two-thirds of new infections occur. It is uncertain whether a vaccine developed for one clade would protect against other clades.

More generally, little research is oriented toward tropical diseases. Pecoul et al. (1999) report that of the 1,233 drugs licensed worldwide between 1975 and 1997, only 13 were for tropical diseases. Two of these were modifications of existing medicines, two were produced for the U.S. military, and five came from veterinary research. Only four were developed by commercial pharmaceutical firms specifically for tropical diseases of humans. (Note, however, that the definition of tropical disease used in their assessment was narrow, and that many of the other drugs licensed in this period were useful in both developing and developed countries.)

III. Failures in the Markets for Vaccines and Vaccine Research

One reason for the paucity of research on vaccines for malaria, tuberculosis, and clades of HIV common in Africa is simply that the countries affected by these diseases are poor, and cannot afford to pay much for vaccines. If this were the only reason, however, there would be no particular reason to target aid expenditures to vaccines or vaccine research, rather than to other goods needed in poor countries, such as food and shelter. In fact, however, distortions in the markets for vaccines lead them to be underconsumed even relative to the incomes of the poor. Even more severe distortions in the research market eliminate incentives for private firms to conduct vaccine research that would be cost-effective for society as a whole, even by the stringent cost-effectiveness standards used to evaluate health interventions in poor countries.

The subsection titled Failures in the Market for Vaccines argues that vaccines are underconsumed and that large public purchases can potentially make both vaccine producers and consumers better off than they would be under monopoly pricing. The next subsection, titled Failures in the Market for Vaccine Research, argues that under current institutions, private returns to research are limited by the ease of designing around patents and by temptations for governments to hold down vaccine prices once vaccines have been developed. The third subsection, titled Social vs. Private Return: Some Quantitative Estimates, reports a rough calculation suggesting that vaccines would be cost-effective health interventions for poor countries at prices 10 or 20 times as much as vaccine developers could hope to realize from their work. Thus, under current institutional arrangements, private developers will lack incentives to pursue socially valuable research opportunities.

Failures in the Market for Vaccines

Vaccines are underconsumed for a variety of reasons. First, individuals who take vaccines not only benefit themselves, but also help break the chain of disease transmission, thus benefiting the rest of the population. Individuals have no incentive to take these external benefits into account in deciding whether to be vaccinated. Second, the chief beneficiaries of vaccines are often children. Even if the cost of vaccination is trivial relative to the extra future wages children will earn if they stay healthy, children cannot contract to pay for vaccination out of those future wages. Third, consumers seem much more willing to pay for treatment than prevention. Many potential consumers in developing countries are illiterate and place limited credence in official pronouncements about the benefits of vaccination. They may wait to see these benefits by observing what happens to neighbors who take vaccines. However, the benefits of vaccines, unlike those of drugs for treating diseases, are difficult to see, since the benefits of vaccines are not evident until considerably after vaccines are taken, and many people who do not take vaccines never get sick.

Monopoly pricing of vaccines would exacerbate underconsumption of vaccines. This may explain why governments in the vast majority of countries purchase vaccines and distribute them to the population either free or at a highly subsidized price. Because vaccine development is expensive, but manufacturing additional doses of vaccine is typically cheap, large government purchases can potentially make both vaccine producers and the general public better off than they would be under monopoly pricing to individuals. This can be achieved by purchasing a large quantity of the vaccine at a lower price per dose than under monopoly pricing to individuals. The vaccine developer can be made better off if the total value of their sales (price times quantity) is higher than it would be under sales to individuals. Those consumers who would have been willing to pay the monopoly price are better off, as long as the taxes they would have to pay to finance government vaccine purchases are less than the monopoly price. The consumers who valued the vaccine at more than the production cost but less than the monopoly price can also be made better off, as long as the value they place on the vaccine is greater than the increase in taxes necessary to finance government purchases.

Figure 2.1 shows a situation in which government purchases can potentially make everyone better off than under monopoly pricing. The downward sloping line shows the willingness to pay of different potential consumers for the vaccine, which depends on their income. The lower horizontal line represents the cost of producing an additional dose of the vaccine once the research costs have been incurred and the

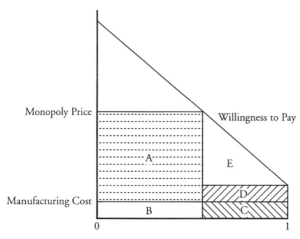

Figure 2.1

factory has been built. A monopolist will choose a price to maximize profits. Area A represents the surplus of revenue over marginal manufacturing costs under monopoly pricing. These funds can be used to cover the costs of research and development on the vaccine, the costs of building the factory, and any profits. Note that many people who are not willing to buy the vaccine at the monopoly price would be willing to pay more than the amount it costs to produce an additional dose of vaccine.

To see why large government purchases that expand the market and bring down the average cost per dose may potentially be able to make everybody better off, suppose that the government agrees to pay the vaccine manufacturer an amount equal to the sum of areas A, B, C, and D in exchange for enough vaccines for the entire population. If these purchases are funded by taxing people based on their income, with all people who would have paid the monopoly price paying just under that price, and all other people paying just over the actual production cost,[3] vaccine producers and the general public will both be better off than under monopoly pricing.[4,5] Areas D and E represent the social benefit of the vaccine purchase program.

Note that while large government purchases could *potentially* make both consumers and producers better off, if governments force prices too low, they risk making vaccine developers worse off than under a private market system, thus discouraging research.

Failures in the Market for Vaccine Research

Economists have estimated that the social returns to research and development are typically twice the returns to private developers (Nadiri 1993; Mansfield et al. 1977). Private developers therefore lack incentives to pursue research on socially valuable projects. The gap between private and social returns to research is likely to be much greater for research on malaria, tuberculosis, and HIV vaccines than in many other areas of applied research. This is because it is often possible to design around vaccine patents and because governments are often tempted to use their powers as regulators and large purchasers to hold down vaccine prices after firms have sunk their research investments and developed a vaccine. Because vaccine research is a global public good benefiting many small countries, no single country has an incentive to pay higher vaccine prices to encourage research.

It is often possible to design around vaccine patents, and this may make it difficult for the original developers of new vaccines to recoup their research expenditures. Once they have invested millions in a risky effort to develop new vaccines, competitors may be able to slightly alter their approach so as to develop a competing vaccine, driving down prices. In many industries, first-mover advantages are often as important as patents in spurring innovation. However, governments and international organizations purchase most vaccines, and these institutions are not particularly subject to brand loyalty.

Vaccine research is subject to what economists call a "time consistency" problem. Vaccine research is very expensive, but once vaccines have been invented, they can usually be manufactured at low cost.[6] Once a vaccine has been developed, even a public-spirited government may be tempted to try to obtain vaccines at a price that would cover manufacturing costs but not research costs. Governments are in a strong bargaining position at this point because they are major vaccine purchasers, they regulate vaccines, and they are arbiters of intellectual property rights. Governments and international organizations therefore can, and do, bargain for very low prices. Thus, while in theory government purchases of vaccines could make both vaccine producers and consumers better off, in practice they are often used as a vehicle to transfer wealth from vaccine producers to consumers.[7] Since potential researchers anticipate this redistribution, they invest less in research than they otherwise would.

The time-consistency problem that leads governments to pay low prices for vaccines is exacerbated by political problems in many developing countries that make vaccines a low political priority. In particular, since vaccines deliver a widely distributed benefit, they tend to receive less political support than expenditures which benefit more concentrated and politically organized groups, including salaries for health workers.

Moreover, vaccine research and development is a global public good, so each country has an incentive to free ride off research financed by other countries' governments or induced by their intellectual property rights protection. A large country, such as the United States, would know that if it did this, it would risk cutting off the flow of future research. Small countries, such as Uganda, can assume that individually their actions will have little effect on total research incentives. However, if all African countries act this way, there will be little incentive for the development of a malaria vaccine.

This free-riding problem is particularly severe for countries that are only a small fraction of the world market and hence reap only a small fraction of the worldwide benefits of research. Pharmaceutical prices are controlled at prices approximately one half of United States levels in the European Union, while in Japan, they are controlled at one quarter of U.S. levels (Robbins and Freeman, 1988). The world's three leading infectious diseases affect many small developing countries that have even less reason to internalize the benefits of drug development than the European Union or Japan.[8]

Historically, developing countries have not provided much protection for intellectual property rights for pharmaceuticals. Until recently, many developing countries did not grant patent protection for pharmaceuticals and thereby kept prices low (Siebeck et. al., 1990). Several developing countries, including India and Brazil, have recently agreed to enhance intellectual property rights for pharmaceuticals, but only under intense trade pressure from the United States. It remains to be seen whether the promised intellectual property rights policies will be enforced. Many pharmaceutical firms are skeptical. The South African government recently announced that it may attempt to force patent holders on AIDS drugs to license their patents to generic manufacturers. The United States initially opposed this, but abandoned its opposition in response to a storm of protest. Given the huge importance of an HIV or malaria vaccine to many developing countries, it is far from clear that the U.S. could induce developing countries to establish strong intellectual property rights for such vaccines, short of offering to pay for the additional costs this would impose on the countries.

Research on vaccines for diseases prevalent in both developed and developing countries has been stimulated by demand in developed countries. However, the limited intellectual property rights available in many poor countries deter research on vaccines against diseases such as malaria, which would have little market in developed countries.

Note that even if intellectual property rights were enforced globally, the same arguments that suggest that national vaccine purchases are more efficient than individual purchases would also suggest that international purchases are potentially more efficient than national purchases. If vaccine developers charge a single monopoly price to governments, some countries will not be able to afford to purchase the vaccine. All countries could potentially be made better off, as long as the rich countries paid no more than the monopoly price they would have paid otherwise, and the poor countries pay less than the amount at which they value the vaccine, but more than the actual production cost. Note also that even if poor countries could somehow be induced to establish strong intellectual property rights for vaccine developers, they would still have market power as purchasers, and hence would still likely be able to negotiate a price below the full social value of vaccines. Hence research and development incentives would likely be too small even in this case.

The market could reach efficient size if vaccine developers charged each nation a separate price based on the maximum amount which they were willing to pay, through a system of tiered pricing. In fact, pharmaceutical firms do charge different prices to different countries. However, opportunities for tiered pricing are limited, partly by the possibility of resale, but primarily by fear of a political backlash in rich countries. Politically, it is difficult for pharmaceutical firms to justify charging much higher prices in one country than in another. For example, after a Congressional hearing in which Senator Paula Hawkins asked a major vaccine manufacturer how it could justify charging nearly three times as much to the United States government for vaccines as to foreign countries, U.S. manufacturers stopped submitting bids to UNICEF to supply vaccines.[9]

One way to achieve some of the same objectives as tiered pricing would be to purchase vaccines internationally for a range of poor countries at a single price and then collect copayments from these countries that would vary with their incomes. This approach would increase access to vaccines, while ensuring that richer countries, which have greater willingness to pay for vaccines, contribute more toward covering the costs of vaccine research and development. The embarrassment of charging many different prices to different countries would be avoided. This approach would, however, require outside funding to make up the difference between the price at which vaccines are purchased from manufacturers and the copayments received from the poorest countries.

Social vs. Private Return: Some Quantitative Estimates

A crude preliminary estimate suggests that the social benefits of vaccines may be 10 to 20 times the private benefits appropriated by vaccine developers. Since potential vaccine developers will consider only private returns in setting their research budgets, incentives for vaccine research are almost certainly far too small.

Consider the potential benefits from a hypothetical 80% effective one-dose malaria vaccine. A standard way to assess the cost-effectiveness of a health intervention is the cost per Disability Adjusted Life Year (DALY) saved. In its 1993 World Development Report, the World Bank defined health interventions as "highly cost effective" for poor countries if they cost less than $100 per DALY saved. (In contrast, health interventions are considered cost effective in the U.S. at up to 500 to 1000 times this amount—$50,000–$100,000 per year of life saved (Neumann et al. 2000).)

The WHO recently estimated that malaria costs 39.3 million DALYs per year (WHO, 1999a). Malaria is particularly deadly in children under five, who have not yet developed limited natural immunity, and women pregnant with their first child, whose immune systems are suppressed. The target population for a malaria vaccine would be the roughly 50 million children born annually in low-income and lower-middle-income countries with high enough prevalence to make vaccination cost effective and the approximately 10 million women pregnant with their first child living in countries with high enough prevalence to make vaccination of this group cost-effective.[10] We assume that 75% of targeted children and 50% of targeted first-time mothers are reached, so that 42.1 million people are immunized annually. Incremental delivery costs for adding a single-dose childhood vaccine to the EPI package might be about $1.00 per child vaccinated.[11] The marginal cost of delivery for pregnant women might be closer to $3.00

per woman vaccinated. A rough calculation suggests that delivery of such a vaccine would save 17.6 million DALYs each year and would cost $52.2 million annually for a delivery cost of about $2.97 per DALY saved. This implies that at a cost-effectiveness threshold of $100 per DALY, an 80% effective malaria vaccine would be cost-effective even at a price of $41 per immunized person, or a total of $1.73 billion annually to immunize 42.1 million people.[12] Note that these figures do not take into account knock-on reductions in secondary infections or the potential economic benefits of reducing malaria prevalence beyond the impact on the individual suffering from the disease.[13] (See Glennerster and Kremer 2000 for more detailed calculations of vaccine cost-effectiveness.)

These calculations imply that from the standpoint of society as a whole, it would be cost effective for private developers to conduct research leading to a malaria vaccine, even if the research were risky and expensive enough that the developer would have to charge $41 per immunized person, or $1.73 billion annually, in perpetuity, to recoup the research costs and the risk of failure. However, such a research investment would not be cost-effective from the standpoint of a private developer. To give some indication of this, the total developing country market for childhood vaccines is $200 million annually (World Bank AIDS Vaccine Task Force 2000). The combined cost of the six vaccines in the standard Expanded Program on Immunization (EPI) package is about $0.50 (Robbins and Freeman, 1988). Of course, a vaccine under patent would likely generate greater revenues than off-patent vaccines. However, when the hepatitis B vaccine was first introduced and priced at $30 per dose, it was used infrequently in developing countries (Muraskin 1995; Galambos 1995).[14] Even at a dollar or two per dose, hepatitis B and Haemophilus influenzae b vaccines do not reach most children in the poorest countries (General Accounting Office 1999). It seems likely that the developer of a malaria vaccine would receive payments worth less than one-tenth or one-twentieth of the $41 per immunized person at which vaccines would be cost-effective. The huge disparity between private incentives to invest in research and the social benefits of a vaccine suggests that research investment will be far too little in the absence of public support.

To summarize, vaccine research is an international public good, since efforts by one country to develop a malaria vaccine will benefit others as well. Once vaccines are developed, governments may be tempted not to compensate vaccine developers for their research expenditures, so potential developers will not invest in research without credible commitments that they will be paid. A rough quantitative estimate suggests that vaccine developers will lack incentives to pursue malaria, tuberculosis, and HIV vaccine research, even if this research would be extremely cost effective for society as a whole. These factors suggest that encouraging vaccine research may be very cost effective relative to existing forms of development assistance which do not particularly target global public goods. The next two sections discuss alternative ways to promote vaccine research.

IV.   The Roles of Push and Pull Programs in Encouraging Vaccine Research

The literature on vaccine research distinguishes between push and pull programs. Push programs provide funding for vaccine research, for example through grants to academics, public equity investments in vaccine development, research and development tax credits, or work in government laboratories. Pull programs increase rewards for development of a vaccine, for example by promising to purchase a vaccine if it is developed. Roughly, the distinction is between paying for research inputs and paying for research outputs. The first subsection, titled The Potential Role of Pull Programs, argues that pull programs are well suited to the later stages of the vaccine development process, and discusses some of the problems with push programs, as illustrated by the history of USAID's push program to develop a malaria vaccine. The second subsection, titled Combining Push and Pull Programs, argues that push programs are well suited to financing basic research, and discusses how push and pull programs can be combined.

The Potential Role of Pull Programs

Historically, programs designed to encourage vaccine research financed research inputs ahead of time rather than offering to pay for a vaccine.[15] This may have been in part because there were relatively few sources of finance for commercial pharmaceutical research outside a few major pharmaceutical companies. However, the rise of the biotech industry, the availability of venture capital, and the increased willingness and ability of large pharmaceutical firms to contract with smaller firms and universities have made it much easier for researchers with reasonable scientific prospects of developing a product to attract outside investors, as long as a sufficient market is expected for the product. Pull programs could create such a market. It is worth reevaluating methods of supporting research in light of this changed institutional environment.

Under pull programs the government pays nothing unless a vaccine is developed. This creates strong incentives for researchers to (1) carefully select research projects and (2) focus on developing viable vaccines, rather than pursuing other goals.

Perhaps the chief advantage of pull programs that provide strong financial incentives for production of a vaccine is that they help in selecting research projects. This is true both at the level of selecting individual research projects and at the level of determining whether a major research effort on vaccines for malaria, tuberculosis, or HIV is scientifically warranted at all. Researchers working on a particular line of research have an interest in exaggerating the promise of their own lines of research. Scientific administrators may have trouble deciding which diseases are worth working on, and which vaccine approaches, if any, are worth pursuing. They may wind up financing ideas with only a minute probability of success, or worse, failing to fund promising vaccine research because they do not have confidence that its backers are presenting objective information on its prospects.

Public sector equity investments in vaccine development projects are subject to a similar problem. Firms that believe they have identified projects with very high expected net present value will be least inclined to seek public sector investments that would dilute their equity stake, while those who are least confident about their research prospects will be most inclined to seek outside equity investment.

Even if government-directed research programs manage to initially select appropriate research projects, they are likely to fail to revise these judgments in light of later evidence. If results on a particular research project that initially appear promising later turn out to be disappointing, a private firm is likely to shut the project down. A public entity may acquire its own bureaucratic momentum, leading governments to throw good money after bad. Public sector institutions are notoriously difficult to shut down.

The problem of selecting research projects exists not only on the level of deciding which research avenues toward a particular vaccine are most promising, but also at the level of deciding whether to expand vaccine research at all. The previous section on Failures in the Market for Vaccine Research argued that since vaccines would be cost-effective at prices much greater than vaccine developers could hope to receive, private developers would have an incentive to pass up research opportunities that were cost effective from the standpoint of society as a whole. This does not prove that such opportunities exist. Elected officials and the public are likely to find it very difficult to assess the scientific opportunities for research on malaria, tuberculosis, and AIDS vaccines. Under a system of grant financed research, advocates for particular diseases and scientists working on the disease have an interest in exaggerating the opportunities.

While the gap between private and social incentives for vaccine development does not prove that socially useful research opportunities exist, it does suggest a case for bringing private and social incentives into line, so that private developers will have incentives to pursue any socially desirable research investments that do exist. A vaccine purchase commitment can do this. Since taxpayers pay nothing unless and until an effective vaccine is produced, elected officials and the public do not have to worry that they are investing millions to develop a technically infeasible vaccine. Government officials do not have to decide between competing scientific approaches. Pharmaceutical firms contemplating pursuing a line of research and scientists contemplating joining biotech ventures in exchange for stock options will invest their money and time only if they believe the scientific prospects are promising. Purchase commitments have an advantage over research grants or equity investments precisely because the scientific potential for vaccines is difficult for outsiders to assess.

In addition to allowing researchers to self-select promising projects, pull programs encourage researchers to focus intently on developing a marketable vaccine, rather than on other goals. Many academic and government researchers have career incentives and intellectual interests that orient them to fundamental science. In contrast, the later, more applied stages of vaccine development include activities that are not particularly interesting intellectually, but are expensive. Techniques for manufacturing sufficient quantities of candidate vaccines in sufficient purity for clinical trials must be developed. Animal models for the disease must be created. Vaccine trials in the field must be conducted. Nobody wins a Nobel Prize for these important steps in vaccine development. By linking payment to results, pull programs provide strong incentives to researchers to concentrate their efforts on development of a vaccine.[16] Under a system of grant financed research, it can be difficult to monitor that researchers are focusing on development of a vaccine rather than publishing articles or applying for the next grant.

A similar monitoring problem arises when private research is subsidized through targeted R&D tax credits. Currently, U.S. companies are eligible for a 20% research and development tax credit. A bill recently introduced in the United States Congress proposes increasing this credit to 30% for research on vaccines for diseases that kill more than one million people a year. One potential problem with such an approach is that firms doing research with only indirect implications for these diseases might try to claim eligibility for the credit, while focusing much of their effort on developing more lucrative products.[17] In contrast, a tax credit linked to the sale of a vaccine, such as that proposed in President Clinton's 2000 budget, does not face the same monitoring problems. This credit will only be awarded if a marketable vaccine is produced.

Another problem with push programs is that when governments directly allocate research spending up front, they sometimes base decisions on political, rather than scientific, considerations. For example, there may be pressure to spend funds in particular congressional districts. The analogue for internationally supported research on malaria, tuberculosis, and HIV is political pressure to allocate research expenditures to particular countries, developing countries in particular. With pull programs, in contrast, the sponsors promise to pay for a viable vaccine wherever it is developed.

The risks that grant funded scientists and research administrators competing for budgets will overestimate the chances of success and divert resources away from vaccine research are far from hypothetical. Desowitz (1991) chronicles the sad story of the U.S. Agency for International Development's 1980s push program to develop a malaria vaccine. USAID's efforts focused on three teams. A candidate vaccine was developed by the first team. Tests with nine volunteers found that only two were protected from malaria, and suggested that the vaccine created side effects. These results, mixed at best, did not prevent USAID from issuing wildly overoptimistic statements. In 1984, the agency claimed that there had been a "major breakthrough in the development of a vaccine against the most deadly form of malaria in human beings. The vaccine should be ready for use around the world, especially in developing countries, within five years."[18] Fifteen years later, the world is still waiting for a malaria vaccine.

Early work by the second team yielded disappointing results, but not surprisingly, the principal investigator argued that his approach was still worth pursuing and requested an additional $2.38 million from USAID. The expert consultants assigned to review the project recommended that the research not be funded. However, USAID's malaria vaccine project director told the USAID Office of Procurement that the expert panel "had endorsed the scientific methodology and the exceptional qualifications and experience of the researchers."[19] Once the grant came through, the principal investigator transferred grant funds to his personal account. He was later indicted for theft.

The external evaluations of the third proposal called it mediocre and unrealistic. The USAID project director ignored the report and arranged for the project to be fully funded. The principal investigator and his administrative assistant were later indicted for theft and criminal conspiracy in diverting money from the grant to their personal accounts. Two months before his arrest, the Rockefeller Foundation had provided him with a $750,000 research grant, and on the very day that he was arrested, USAID announced it was giving him an additional $1.65 million for research.

By 1986, USAID had spent over $60 million on its malaria vaccine efforts, with little progress. Since USAID believed that there would soon be many candidate malaria vaccines suitable for testing, it tried to obtain monkeys as test subjects for these vaccines. USAID's malaria vaccine project director, James Erickson, arranged for a contract to acquire monkeys to go to an associate who paid him a kickback. Erickson eventually pleaded guilty to accepting an illegal gratuity, filing false tax returns, and making false statements.

What about outside oversight? USAID had arranged for independent oversight to be provided by the American Institute of Biological Science (AIBS). Erickson and the AIBS-assigned project manager were lovers.

The USAID case is extreme, and many push programs are quite successful. But the general principle remains that researchers funded under push programs have incentives to be overoptimistic, and since they are paid before delivering a product, they may be tempted to divert resources away from the search for a vaccine.

The USAID example may shed light on why the administrators of push programs and the researchers financed by such programs often believe that push programs are somehow cheaper than pull programs.

As a first approximation, a biotech or pharmaceutical firm will find it profitable to take on a project if the probability of success times the net present value of profits if the project succeeds exceeds the cost of undertaking the project. This implies that even in the best case, if the government funds only worthwhile research projects and researchers focus all their energies on developing a vaccine, the expected discounted cost of developing a vaccine is likely to be similar in net present value terms whether research is financed at the front end, through government grants; or induced by payments for a successful vaccine at the back end.[20] In the more likely case, when research organizations are more careful in selecting projects and more focused on developing vaccines if they are only paid if they succeed, private research is likely to be more cost-effective than government programs.

Why then do many government scientists argue that push programs are cheaper than pull programs? The USAID example illustrates that researchers are prone to underestimate their costs and overestimate their chance of success. Of course, scientists in pharmaceutical firms do the same. But pharmaceutical executives and biotech investors anticipate this overoptimism, and correct for it by requiring high projected hurdle rates before approving projects or investing funds. The net effect is that pharmaceutical executives and biotech investors wind up approving projects that are likely to have positive net present value after correcting for the overoptimism of project proponents. It is misleading to compare the amounts government scientists claim they would need to develop vaccines with the markets pharmaceutical executives claim they would need to justify vaccine investments.

The analysis above is theoretical, but it is consistent with the empirical evidence, which suggests that both government and private R&D have strong positive returns, but that the rate of return on private R&D is substantially greater (Nadiri 1993; Nadiri and Mamuneas 1994; and Bernstein and Nadiri 1988, 1991). The empirical record of government efforts to pick winners in research and development of commercial products is littered with failures, from supersonic transport to the breeder reactor to the Carter oil shale program.

In summary, while the case of USAID's malaria vaccine program is extreme, and many push programs are effective, push programs in general are vulnerable to overoptimism and monitoring problems. With pull programs, in contrast, biotech and pharmaceutical firms spend their own money on research, and the public pays only if a vaccine is produced.

Combining Push and Pull Programs

Although pull programs have an advantage in the later stages of development, push programs are likely to be well suited to financing basic research. The main objective of basic research, by definition, is to provide information to other researchers rather than to develop products. A program that ties incentives to the development of a product would encourage researchers to keep their research results private as long as possible in order to have an advantage in the next stage of research. In contrast, grant funded academics and scientists in government laboratories have career incentives to publish their results quickly. (One way around this would be to link payment to research output, but it is difficult to measure the quality of basic research. It is not simply a matter of testing if a vaccine works, or a product sells.)

Push programs also have some attractive features even for later stages of research. To the extent that intermediate steps in the vaccine research and development process create spillovers for other researchers, it might be worth considering providing milestone payments tied to these intermediate steps. For example, milestone payments could be paid if efficacy were demonstrated in animals. However, milestone payments do not target the ultimate objective of a vaccine, and hence might stimulate wasteful investments in research lines that were unlikely to lead to a viable vaccine. For example, researchers might try to demonstrate efficacy in animal models for a vaccine that was unlikely to be safe in humans. This problem is greater the larger the milestone payment; if a milestone payment is greater than the cost of performing the research, firms might find it profitable to reach the milestone even if they know they can go no further. Milestone payments will be less likely to stimulate wasteful research on candidates unlikely to yield a viable vaccine if they are given in the form of subsidies for future research on the candidate vaccine.

It is not clear whether the same body which administers a vaccine purchase commitment program should also award milestone payments. On the one hand, a track record of milestone payments could help build credibility for a vaccine purchase commitment. On the other hand, a committee that had supported, or not supported, a line of research through milestone payments might find it difficult to be objective in assessing eligibility and pricing for a vaccine purchase commitment.

Push programs have other advantages. Government programs that pay for research whether it succeeds or fails transfer the risk of failure from the research firms' shareholders to society at large, and to the extent that shareholders cannot diversify risk in the stock market, this risk spreading is a potential advantage of push programs. A number of theoretical models suggest that private firms competing for a patent may inefficiently duplicate each other's activities. A centralized program may prevent this. (On the other hand, while decentralization may lead to some duplication of effort, it also means that mistakes by a single decision maker will not block progress toward a vaccine.)

One of the biggest advantages of push programs relative to pull programs (other than patents) is that they do not require specifying the output ahead of time. A pull program could not have been used to encourage the development of the Post-it Note or the graphical user interface, because these products could not have been adequately described before they were invented. In contrast, it is comparatively easier to define what is meant by a safe and efficacious vaccine, and existing institutions, such as the U.S. FDA, are already charged with making these determinations. As discussed in the companion paper, "Creating Markets for New Vaccines: Design Issues," even for vaccines, however, defining eligibility standards is far from trivial.

In general, society seems to prefer to use direct government support for basic research, while using the promise of an exclusive market, rather than centralized government programs, to stimulate the applied work of actual product development. Applying the same principle to vaccines would suggest using the promise of a market to encourage applied vaccine research.

Some push programs are already in place to spur vaccine research, although funding is modest. For example, the International AIDS Vaccine Initiative (IAVI) supports AIDS vaccine efforts. In contrast, there are currently no programs in place to fully reward developers of viable malaria, tuberculosis, or HIV vaccines. If the already existing push programs were complemented with pull programs, researchers would still have an incentive to pursue any promising research leads that slip through the cracks of the push system.

If vaccine research were supported through a mix of push and pull programs, push funders could insist on a share of revenues if a project they support leads to a vaccine that is rewarded through a pull program, or could condition public financing on agreement to supply the vaccine to poor countries at a modest markup over manufacturing costs.

V. Alternative Pull Programs

Pull programs that reward successful vaccine research could take several different forms other than commitments to purchase vaccines, including extensions of patent rights on other products, cash prizes, research tournaments, and signaling willingness to pay more for future vaccines by purchasing more existing vaccines at a higher price.[21] Given the huge disparities between private and social returns to research, it is likely that any reasonable program to reward vaccine developers would be cost-effective relative to the alternative of sticking with the status quo. However, this section argues that extensions of patent rights on other pharmaceuticals are not the most efficient way to reward vaccine developers; that while cash prizes and commitments to purchase vaccines are economically quite similar, purchase commitments are likely to be somewhat more attractive politically, and thus more credible to potential vaccine developers; and that research tournaments are inappropriate for situations like vaccine development, in which it is possible that no satisfactory product will be created by a given date. Purchasing and distributing currently underutilized vaccines is certainly justified in its own right, but on its own is unlikely to convince potential developers of vaccines for malaria, tuberculosis, or African strains of HIV that the international community will be willing to pay for these vaccines in 10 or 15 years.

Patent Extensions

Jonathan Mann, the late founding director of the WHO Global Program on AIDS, suggested compensating the developer of an HIV vaccine with a 10-year extension of patent rights on another pharmaceutical. With successful pharmaceuticals bringing in as much as $3.6 billion in annual sales (CNNfn 1998) such a patent extension would be very valuable. Patent extensions may be politically appealing to advocates, in that they need not go through the budget process. However, they inefficiently and inequitably place the entire burden of financing vaccine development on patients in need of the drug for which the patent has been extended. To see this, note that extending the patent on Prozac as compensation for developing an HIV vaccine is economically equivalent to imposing a high tax on Prozac and using the proceeds to finance cash compensation for the HIV vaccine developer. High taxes on narrow bases are typically an inefficient way of raising revenue, since they distort consumption away from the taxed good.[22] An extension of the Prozac patent would prevent some people from getting needed treatment for depression.

The potential countervailing advantage of patents is that when they are applied to the invented good, they closely link the inventor's compensation to the value of the invention, since inventors will be able to charge more for valuable inventions. If a vaccine is more effective, causes fewer side effects, and is easier to administer, it will bring in more revenue. Patents therefore create appropriate incentives for potential inventors. However, rewarding the inventor of an HIV vaccine with the extension of a Prozac patent eliminates this link between the usefulness of the invention and the magnitude of the compensation.

Another disadvantage of compensating vaccine inventors with extensions of patents on unrelated pharmaceuticals is that the right to extend a patent would be worth the most to firms holding patents on commercially valuable pharmaceuticals, and these firms may not be those with the best opportunities for vaccine research. This problem would not be fully resolved by making patent extensions tradable, since firms holding patents on commercially valuable pharmaceuticals would presumably receive some profits in any such trades. If vaccine developers were compensated in cash, rather than patent extensions, they could receive the full value of the compensation without sharing it with the holders of patents on unrelated pharmaceuticals.

Cash Prizes

Cash prizes in lieu of patents are economically similar to purchase commitments. However, purchase commitments more closely link payments to vaccine quality and are more politically attractive, and hence more credible. The disadvantages of government purchases are likely to be minor for vaccines.

Compared to cash prizes in lieu of patents, vaccine purchases provide a closer link between payments and vaccine quality. For example, suppose that a vaccine received regulatory approval, but was later found to have side effects. If a cash prize had been awarded at the date of regulatory approval, it might be difficult to get the money back. Vaccine purchases, on the other hand, could be suspended if countries wished to cease purchasing vaccines.

Moreover, purchase commitments are likely to be politically more attractive than cash prizes, and thus more credible to potential vaccine developers. Vaccine developers are vulnerable to expropriation, even if the terms of the compensation program legally obligate the government to provide compensation for any qualifying vaccine: the funds could be extracted from them in a supposedly separate, unrelated action. For example, a pharmaceutical firm that had just earned a windfall on a malaria vaccine might be subject to stiff price regulation on another product. This suggests that it is important to design a compensation program in ways that are as politically acceptable as possible, and that generate the minimum amount of resentment. Purchasing malaria vaccine for the 50 million children born in Africa each year at $5 a dose for 10 years is likely to be more politically appealing than awarding a $2.5 billion prize to a pharmaceutical manufacturer. Conversations with pharmaceutical executives suggest that they do not like anything labeled as a prize.

Cash prizes in lieu of patents lead to free competition in manufacturing newly invented goods, whereas public purchases require the government to specify details of the goods purchased. This would represent a significant advantage of prizes over purchases for most goods, but it is less important for vaccines. For example, if the government committed to purchase high definition television sets as a way of encouraging research, it would have to get involved in decisions about screen size, color, style, reliability, and other issues best left to consumers. In contrast, governments regulate vaccine quality in any case. Moreover, an effective malaria vaccine would be easy to allocate, since a single course would presumably be taken by all children in malarious areas.[23]

Tournaments

In research tournaments, the sponsor promises a reward to whoever has progressed the farthest in research by a certain date. (See Taylor 1995 for a discussion of tournaments.) The design competitions often used to select architectural firms are examples of tournaments. In a vaccine tournament, a committee might be established with instructions to award a cash prize to whichever research team had made the most progress toward a vaccine as of a specific date. If no vaccine had been completed by that date, additional funds could be set aside for further rounds of the tournament.

Tournaments have several limitations, however, and may not be appropriate for encouraging vaccine research.

First, a payment must be made no matter what is developed. While tournaments provide incentives for researchers to devote effort to developing a product, they do not address the problem of determining whether research on a particular vaccine is worth pursuing at all. Advocates for a particular disease and scientists working on the disease will always want to encourage the establishment of tournaments for research on their disease, even if the prospects for ultimate success are low. With a vaccine purchase program, nothing is spent unless a vaccine is developed.

Another problem with tournaments is that once research has been completed, the award committee might be tempted to allocate the reward on grounds other than progress in research. The committee might award the reward to a more politically correct firm, to a university team, or to whoever had done the most scientifically interesting work, rather than to the team which had made the most progress toward a vaccine. Anticipating this, firms might invest in political correctness or scientific faddishness rather than in producing an effective vaccine. Of course, a committee making purchase decisions for a vaccine purchase program could also be subject to bias, but judgments about who has made the most progress developing a vaccine are more subjective than judgments about whether a vaccine with a particular set of results from phase III trials is satisfactory.

Collusion among potential researchers may be particularly harmful in tournaments. If only a few pharmaceutical firms had done a significant amount of work, they could collude to exert low effort on doing further research, since the reward would be paid whether or not a vaccine was developed.

Tournaments may lead researchers to put their efforts into looking good on the tournament completion date, rather than completing a vaccine. Firms which discovered promising research leads that were unlikely to yield solid results before the deadline might ignore their leads, while firms that received information that the research line they were pursuing would not yield a vaccine might not reveal this information.

Tournaments are also politically unattractive. Governments may not find it politically attractive to pay large amounts for research that may have not progressed very far. Since there would be no clear-cut way to decide who was ahead in research, awards might be subject to litigation and charges of favoritism.

Finally, rewards in tournaments would have to be in cash, rather than in guaranteed sales, since no vaccine may have been developed by the end of the tournament. As noted earlier, however, cash rewards are less politically attractive than guaranteed markets.

Expanding the Market for Existing Vaccines

Some argue that by purchasing more existing vaccines at higher prices, policymakers can signal their intention to provide a market for future vaccines, and thus encourage research on new vaccines. Although the standard EPI package of vaccines is widely distributed, a number of effective vaccines that are already available are not fully used.[24] Purchasing and distributing existing vaccines which are not widely used in developing countries, such as *Haemophilus influenzae* type b (Hib) vaccine, would be a cost-effective way to save many lives.

However, by itself, paying more for currently available vaccines may not make pharmaceutical firms confident that they will be rewarded for developing new vaccines. It could easily take 10 years to develop malaria, tuberculosis, or HIV vaccines, and developers would need to recoup their investment through sales in the 10 years following the vaccines' development. Since international interest in health in developing countries is fickle, pharmaceutical firms might well feel that the availability of funds to purchase Hib vaccine now at a remunerative price does not guarantee that the international community would be prepared to pay much for future vaccines 15 years from now. Legally binding commitments to purchase future vaccines at specified prices would still play a critical role in spurring research.

Moreover, given that the Hib vaccine was developed without any expectation of realizing substantial profits in developing countries, paying more than pharmaceutical firms could reasonably have expected for these vaccines would provide extra profits to pharmaceutical firms. Providing these extra profits might be worthwhile if it were the only way to establish a reputation for paying remunerative prices for future vaccines. Not surprisingly, pharmaceutical manufacturers argue that the best way to persuade them that work on future vaccines would be rewarded would be to buy currently available vaccines at a high price. However, if it were possible to commit now to purchase future vaccines at a remunerative price, there would be no reason to pay more for current vaccines than had been implicitly or explicitly promised to vaccine developers. Paying high prices for both current and future vaccines as a way of encouraging future research amounts to paying twice.

Finally, some argue that increasing current vaccine sales will increase vaccine R&D budgets because pharmaceutical firms finance research on a division by division basis, as a percentage of current sales. It is possible that some firms might use such a rule of thumb to reduce unproductive competition for funds among divisions seeking to increase their R&D budgets. While some pharmaceutical firms may find this rough rule of thumb useful under the current environment, if the environment changes, they will have incentives to change these rules. In particular, if there is an explicit, credible commitment to purchase vaccines, there is reason to think that companies would change their R&D budgeting rules. Finally, note that even if some firms are particularly subject to wasteful internal budget battles and therefore impose draconian internal budget rules, there will be even greater incentives for other firms to expand R&D and for new biotech firms to enter the field in response to increased markets.

In summary, increased purchases and delivery of existing vaccines are likely to be very cost-effective ways of saving lives in their own right. However, in order to motivate R&D on future vaccines, it is necessary to supplement increased purchases of existing vaccines with explicit commitments to reward developers of future vaccines. Paying more for existing vaccines than vaccine developers could have reasonably expected when they invested in research is likely to be an expensive way of encouraging research on future vaccines.

VI. Conclusion

This paper has argued that private incentives for research on vaccines for malaria, tuberculosis, and strains of HIV common in Africa are likely to be a small fraction of the social value of new vaccines, so that under current institutions, potential vaccine developers would have incentives to pass up socially valuable research opportunities. Moreover, if vaccines were developed, access would be limited if they were sold at monopoly prices.

Commitments to purchase vaccines and make them available to developing countries for modest copayments could both provide incentives for development of vaccines, and ensure that vaccines reach those who need them. Taxpayers would pay only if a vaccine were developed.

A companion paper, "Creating Markets for New Vaccines: Part II: Design Issues," discusses the design of these vaccine purchase commitments.

Notes

I am grateful to Daron Acemoglu, Philippe Aghion, Martha Ainsworth, Susan Athey, Abhijit Banerjee, Amie Batson, David Cutler, Sara Ellison, Sarah England, John Gallup, Chandresh Harjivan, Eugene Kandel, Jenny Lanjouw, Sendhil Mullainathan, Ariel Pakes, Ok Pannenborg, Sydney Rosen, Andrew Segal, Scott Stern, and especially Amir Attaran, Rachel Glennerster, and Jeffrey Sachs for very extensive comments. Amar Hamoudi, Jane

Kim, and Margaret Ronald provided excellent research assistance. This paper is part of a Harvard Center for International Development project on vaccines. I thank the National Science Foundation and the MacArthur Foundation for financial support. The views expressed in this paper are my own, and not theirs. Department of Economics, Littauer 207, Harvard University, Cambridge, MA 02138; mkremer@fas.harvard.edu.

1. The vaccine has been much more effective in some trials than others: trials in Britain suggest effectiveness up to 80%, while those in the southern United States and southern India suggest close to zero effectiveness.

2. Vaccination rates are uneven around the world, but the 74% worldwide vaccination rate does not just reflect rich country experience: of the 118 million children born each year, 107 million are born in developing countries.

3. Pharmaceutical manufacturers may try to sell the vaccine to different customers at different prices. However, the ability of pharmaceutical manufacturers to discriminate between customers in this way is limited, because all customers will try to obtain the vaccine at the lower price. The government has the power to tax higher income earners at a higher rate. Pharmaceutical manufacturers may come up with crude income indicators, for example by selling at a discount to groups of hospitals, but they have less scope to vary prices with income than the government does to vary taxes with income.

4. Note that if the willingness to pay for vaccines depends on factors other than income, then tax-financed government vaccine purchases may not make literally everyone better off, because some people may not want to take the vaccine at any price. To see this, it is useful to consider the cases of malaria and HIV. If a safe, cheap, and effective malaria vaccine were developed, almost everyone living in areas with malaria would presumably want to purchase it. On the other hand, some people might not want to take an AIDS vaccine, even if it were free, because they believe that they have a very low chance of contracting the disease. Since taxes would presumably fall equally on people with a low and a high risk of contracting AIDS, large government purchases of an AIDS vaccine might not literally make everyone better off. The willingness of people in low-risk groups to pay for the vaccine might be less than the increase in their taxes necessary to pay for vaccine purchases.

5. As discussed in a companion paper, "Creating Markets for New Vaccines: Part II: Design Issues," government purchase and distribution of products with large development costs but low manufacturing costs involves its own difficulties. Hence, governments do not purchase and distribute all such products. However, purchasing vaccines is likely to be much easier than purchasing other goods, such as CDs. It is difficult for the government to specify what characteristics a CD would need to be eligible for purchase, or how much to pay CD producers as a function of CD quality. Specifying eligibility and pricing rules for vaccines is easier, albeit far from trivial.

6. Note, however, that new vaccines, particularly those based on conjugate technology, are likely to have somewhat greater manufacturing costs than traditional vaccines.

7. Large liability awards can also be interpreted as a way that governments extract resources from vaccine developers.

8. Data on the distribution of burden of disease by country is limited, but some rough calculations suggest that the share of the worldwide disease burden in the country with the greatest burden ranges from 14% and 18% for HIV and malaria respectively, which disproportionately affect Africa, to 25% for tuberculosis, which is a big problem in India. The share of burden borne by the top four countries is in the 40–50% range.

9. When President Clinton announced his childhood immunization initiative in 1993, he said, "I cannot believe that anyone seriously believes that America should manufacture vaccines for the world, sell them cheaper in foreign countries, and immunize fewer kids as a percentage of the population than any nation in this hemisphere but Bolivia and Haiti." [Mitchell, Philipose, and Sanford, 1993].

10. Existing cohorts of children younger than five might also be vaccinated, but since this is a one-time occurrence, it is ignored in this calculation.

11. The addition of both the hepatitis B and the yellow fever vaccines (which are relatively expensive) to the WHO's Expanded Program of Immunization increased the $15 cost of the program by 15%, or $2.25, including both manufacturing and distribution costs.

12. To see this, note that 17.6 million DALYs $\times$ \$100/DALY = \$1.76 billion = \$52.2 million in delivery costs + \$41 per dose $\times$ 42.1 million doses.

13. Gallup and Sachs (2000) use a cross-country regression approach to estimate that countries with severe malaria grew 1.3% less per year than those without malaria. It is difficult to know the portion of this statistical relationship that is causal.

14. Even if the entire pharmaceutical budget in many African countries went to malaria vaccines, the benefit to a vaccine developer would be far less than the social benefit.

15. Several vaccines were therefore developed primarily in the public sector, and only later licensed out to the private sector for production. For example, the meningococcal meningitis vaccine was developed almost entirely at the Walter Reed Army Institute of Research, and a hepatitis B vaccine was designed by the Hepatitis B Task Force (Muraskin 1995). However, it is not clear that the development of these vaccines in the public sector reflects so much the suitability of the public sector for this task as the barriers facing private sector vaccine development.

16. Of course, to the extent that some of the work required to produce a vaccine is not so intellectually interesting, scientists will need to be paid more to conduct this work [See Stern, 2000].

17. Another problem with the particular form of the research and development tax credit used in the United States is that it rewards incremental R&D spending, thus creating a ratchet effect which limits the rewards for sustained high R&D expenditures [Hall, 1993].

18. From Desowitz 1991, p. 255.

19. From Desowitz 1991, p. 258.

20. The cost of capital may be lower for the government than for pharmaceutical firms, but the difference is not that large.

21. In a previous paper (Kremer 1998), I discuss the possibility of buying out patents, using an auction to establish the patent's value. This can be seen as a method of determining the appropriate cash prize in lieu of a patent. One advantage of this approach is that it can be used even for inventions such as the Post-It note, which could not be defined ahead of time and for which it would be very hard to create even a semi-objective procedure for valuing. On the other hand, the auction procedure for valuing patents described in that paper may be subject to collusion. For products such as vaccines, which are comparatively easier to define ahead of time and for which it is comparatively easy to evaluate effectiveness, advance purchase commitments may be just as effective as patent buyouts, and less subject to collusion.

22. As Michael Rothchild has pointed out to me, if governments and Health Maintenance Organizations (HMOs) purchase pharmaceuticals, patents may be equivalent to a broad-based tax. Nonetheless, patents may still be distortionary if HMOs and governments respond to pharmaceutical prices in their treatment decisions. Governments are less likely to do so than HMOs, and so patent extensions are more attractive in countries with centralized health systems.

23. Note that many people are likely to live in areas where taking the vaccine is a borderline decision, and even in these areas, the appropriateness of vaccination depends primarily on technical issues rather than personal preferences.

24. For example, the hepatitis B vaccine is underused. An effective vaccine for malaria or one of the other major killers would likely be consumed much more widely than the hepatitis B vaccine, since the disease burden of hepatitis B is small relative to that of AIDS, tuberculosis, or malaria. Moreover, malaria kills young children very quickly after infection and the onset of symptoms, whereas hepatitis B infection can remain asymptomatic for decades, and many people may not understand its relation to the deaths it causes from primary hepatic cancer in middle age or beyond.

References

Ainsworth, Martha, Amie Batson, and Sandra Rosenhouse. 1999. "Accelerating an AIDS Vaccine for Developing Countries: Issues and Options for the World Bank." World Bank, Washington, DC.

Batson, Amie. 1998. "Win-Win Interactions Between the Public and Private Sectors." *Nature Medicine* 4(Supp.):487–91.

Bernstein, J., and M. I. Nadiri. 1988. "Interindustry R&D, Rates of Return, and Production in High-Tech Industries." *American Economic Review* LXXVIII:429–34.

Bernstein, J., and M. I. Nadiri. 1991, January. "Product Demand, Cost of Production, Spillovers, and the Social Rate of Return to R&D." Working Paper no. 3625, National Bureau of Economic Research, Cambridge, MA.

Bishai, D., M. Lin, et al. 1999, June 2. "The Global Demand for AIDS Vaccines." Presented at the 2nd International Health Economics Association Meeting, Rotterdam.

Chima, Reginald, and Anne Mills. 1998, June. "Estimating the Economic Impact of Malaria in Sub-Saharan Africa: A Review of the Empirical Evidence." Working Paper, London School of Hygiene and Tropical Medicine.

CNNfn. 1998, May 1. "Merck Slashes Zocor Price."

Cowman, A. F. 1995, December. "Mechanisms of Drug Resistance in Malaria." *Australian and New Zealand Journal of Medicine (Australia)* 25(6):837–44.

Department for International Development (DFID). 2000, October 19. "Harness Globalisation to Provide New Drugs and Vaccines for the Poor, Short Says." Press Release, London.

Desowitz, Robert S. 1991. *The Malaria Capers: Tales of Parasites and People.* New York: W. W. Norton.

Desowitz, Robert S. 1997. *Who Gave Pinta to the Santa Maria? Torrid Diseases in a Temperate World.* New York: W.W. Norton.

DiMasi, Joseph, et al. 1991, July. "Cost of Innovation in the Pharmaceutical Industry." *Journal of Health Economics* 10(2)107–42.

Dupuy, J. M., and L. Freidel. 1990. "Viewpoint: Lag Between Discovery and Production of New Vaccines for the Developing World." *Lancet* 336:733–4.

Financial Times. 2000, February 2. "Discovering Medicines for the Poor."

Galambos, Louis. 1995. *Networks of Innovation: Vaccine Development at Merck, Sharp & Dohme, and Mulford, 1895–1995.* New York: Cambridge University Press.

Gallup, John, and Jeffrey Sachs. 2000, October. "The Economic Burden of Malaria." Working Paper, Harvard Institute for International Development, Cambridge, MA. Downloadable from http://www.hiid.harvard.edu.

GAO (General Accounting Office (U.S.)). 1999. "Global Health: Factors Contributing to Low Vaccination Rates in Developing Countries." Washington, DC.

Glennerster, Rachel, and Michael Kremer. 2000. "Preliminary Cost-Effectiveness Estimates for a Vaccine Purchase Program." Working Paper, Harvard University, Cambridge, MA.

Guell, Robert C., and Marvin Fischbaum. 1995. "Toward Allocative Efficiency in the Prescription Drug Industry," *The Milbank Quarterly* 73:213–29.

Hall, Andrew J., et al. 1993. "Cost-Effectiveness of Hepatitis B Vaccine in The Gambia." *Transactions of the Royal Society of Tropical Medicine and Hygiene* 87:333–6.

Hall, Bronwyn. 1993. "R&D Tax Policy During the Eighties: Success or Failure?" *Tax Policy and the Economy* 7:1–36.

Hoffman, Stephen L., ed. 1996. *Malaria Vaccine Development: A Multi-immune Response Approach.* Washington, DC: American Society for Microbiology.

Institute of Medicine (U.S.). 1991. Committee for the Study on Malaria Prevention and Control: Status Review and Alternative Strategies. *Malaria: Obstacles and Opportunities: A Report of the Committee for the Study on Malaria Prevention and Control: Status Review and Alternative Strategies, Division of International Health, Institute of Medicine.* Washington, DC: National Academy Press.

Institute of Medicine (U.S.). 1986. Committee on Issues and Priorities for New Vaccine Development. *New Vaccine Development, Establishing Priorities, vol. 2: Diseases of Importance in Developing Countries.* Washington, DC: National Academy Press.

Johnston, Mark, and Richard Zeckhauser. 1991, July. "The Australian Pharmaceutical Subsidy Gambit: Transmitting Deadweight Loss and Oligopoly Rents to Consumer Surplus." Working Paper no. 3783, National Bureau of Economic Research, Cambridge, MA.

Kim-Farley, R., and the Expanded Programme on Immunization Team. 1992. "Global Immunization." *Annual Review of Public Health* 13: 223–37.

Kremer, Michael. 1998, November. "Patent Buyouts: A Mechanism for Encouraging Innovation." *Quarterly Journal of Economics* 113(4):1137–67.

Kremer, Michael, and Jeffrey Sachs. 1999, May 5. "A Cure for Indifference." *The Financial Times.*

Lanjouw, Jean O., and Iain Cockburn. 1999. "New Pills for Poor People?: Empirical Evidence After GATT." New Haven, CT: Yale University.

Lichtmann, Douglas G. 1997, Fall. "Pricing Prozac: Why the Government Should Subsidize the Purchase of Patented Pharmaceuticals." *Harvard Journal of Law and Technology* 11(1):123–39.

Mansfield, Edwin, et al. 1977. *The Production and Application of New Industrial Technology.* New York: W. W. Norton & Company.

McQuillan, Lawrence. 1999, September 22. "U.S. Vows to U.N. to Make Vaccines More Affordable." *USA Today:* 6A.

Mercer Management Consulting. 1998. "HIV Vaccine Industry Study October–December 1998" World Bank Task Force on Accelerating the Development of an HIV/AIDS Vaccine for Developing Countries. World Bank, Washington, DC.

Milstien, Julie B., and Amie Batson. 1994. "Accelerating Availability of New Vaccines: Role of the International Community." Global Programme for Vaccines and Immunization. Available at http://www.who.int/gpv-supqual/accelavail.htm.

Mitchell, Violaine S., Nalini M. Philipose, and Jay P. Sanford. 1993. *The Children's Vaccine Initiative: Achieving the Vision.* Washington, DC: National Academy Press.

Morantz, Alison. 2000, February 11. "Preliminary Research on Legal Enforceability of Vaccine Project." Memo, Harvard University, Cambridge, MA.

Muraskin, William A. 1995. *The War Against Hepatitis B: A History of the International Task Force on Hepatitis B Immunization.* Philadelphia: University of Pennsylvania Press.

Murray, Christopher J. L., and Alan D. Lopez. 1996. *The Global Burden of Disease: A Comprehensive Assessment of Mortality and Disability from Diseases, Injuries, and Risk Factors in 1990 and Projected to 2020. Global Burden of Disease and Injury Series; vol. 1.* Cambridge, MA: Published by the Harvard School of Public Health on behalf of the World Health Organization and the World Bank; Distributed by Harvard University Press.

Murray, Christopher J. L., and Alan D. Lopez. 1996. *Global Health Statistics: A Compendium of Incidence, Prevalence, and Mortality Estimates for Over 200 Conditions. Global Burden of Disease and Injury Series; vol. 2.* Cambridge, MA: Published by the Harvard School of Public Health on behalf of the World Health Organization and the World Bank; Distributed by Harvard University Press.

Nadiri, M. Ishaq. 1993. "Innovations and Technological Spillovers." Working Paper no. W4423, National Bureau of Economic Research, Cambridge, MA.

Nadiri, M. Ishaq, and Theofanis P. Mamuneas. 1994, February. "The Effects of Public Infrastructure and R&D Capital on the Cost Structure and Performance of U.S. Manufacturing Industries." *Review of Economics and Statistics* LXXVI:22–37.

National Academy of Sciences. 1996. "Vaccines Against Malaria: Hope in a Gathering Storm." National Academy of Sciences Report. Downloadable from http://www.nap.edu.

National Institutes of Health. 1999. Institutes and Offices. Office of the Director. Office of Financial Management. Funding. http://www4.od.nih.gov/ofm/diseases/ index.stm.

Neumann, Peter J., Eileen Sandberg, Chaim A. Bell, Patricia W. Stone, and Richard H. Chapman. 2000, March–April. "Are Pharmaceuticals Cost-Effective? A Review of the Evidence." *Health Affairs.* pp. 92–109.

Nichter, Mark. 1982. "Vaccinations in the Third World: A Consideration of Community Demand." *Social Science and Medicine* 41(5):617–32.

PATH (Program for Appropriate Technology in Health). At http://www.path.org.

Pecoul, Bernard, Pierre Chirac, Patrice Trouiller, and Jacques Pinel. 1999, January 27. "Access to Essential Drugs in Poor Countries: A Lost Battle?" *Journal of the American Medical Association* 281(4):361–7.

PhRMA. 1999. PhRMA Industry Profile 1999. Available at http://www.phrma.org/publications/industry/profile99/index.html.

Robbins, Anthony, and Phyllis Freeman. 1988, November. "Obstacles to Developing Vaccines for the Third World." *Scientific American* 126–33.

Rogerson, William P. 1994, Fall. "Economic Incentives and the Defense Procurement Process." *Journal of Economic Perspectives* 8(4):65–90.

Rosenhouse, S. 1999. "Preliminary Ideas on Mechanisms to Accelerate the Development of an HIV/AIDS Vaccine for Developing Countries." Technical Paper, The World Bank, Washington, DC.

Russell, Philip K. 1997, September. "Economic Obstacles to the Optimal Utilization of an AIDS Vaccine." *Journal of the International Association of Physicians in AIDS Care.*

Russell, Philip K., et al. 1996. *Vaccines Against Malaria: Hope in a Gathering Storm.* Washington, DC: National Academy Press.

Russell, Philip K. 1998. "Mobilizing Political Will for the Development of a Safe, Effective and Affordable HIV Vaccine." NCIH Conference on Research in AIDS.

Sachs, Jeffrey. 1999. "Sachs on Development: Helping the World's Poorest." *The Economist* 352(8132):17–20.

Salkever, David S., and Richard G. Frank. 1995. "Economic Issues in Vaccine Purchase Arrangements." Working Paper no. 5248, National Bureau of Economic Research, Cambridge, MA.

Scotchmer, Suzanne. 1999, Summer. "On the Optimality of the Patent Renewal System." *Rand Journal of Economics* 30(2):181–96.

Shavell, Steven, and Tanguy van Ypserle. 1998. "Rewards versus Intellectual Property Rights." Cambridge, MA: Harvard Law School. Mimeo.

Shepard, D. S., et al. 1991. "The Economic Cost of Malaria in Africa." *Tropical Medicine and Parasitology* 42:199–203.

Siebeck, W. (ed.), R. Evenson, W. Lesser, and C. A. Primo Braga. 1990. "Strengthening Protection of Intellectual Property in Developing Countries: A Survey of the Literature." World Bank Discussion Paper #112.

Silverstein, Ken. 1999, July 19. "Millions for Viagra, Pennies for Diseases of the Poor." *The Nation* 269(3):13–9.

Sobel, Dava. 1995. *Longitude.* New York: Walker and Company.

Stern, Scott. 2000. "Do Scientists Pay to Be Scientists?" Working Paper no. 7410, National Bureau of Economic Research, Cambridge, MA.

Targett, G. A. T. ed. 1991. *Malaria: Waiting for the Vaccine. London School of Hygiene and Tropical Medicine First Annual Public Health Forum*. New York: John Wiley and Sons.

Taylor, Curtis R. 1995, September. "Digging for Golden Carrots: An Analysis of Research Tournaments." *The American Economic Review* 85:872–90.

UNAIDS. 1998, December. *AIDS Epidemic Update*. Geneva.

Wellcome Trust. 1996. *An Audit of International Activity in Malaria Research*. Downloadable from www.wellcome.ac.uk/en/1/biosfginttrpiam.html.

WHO (World Health Organization) 1999a. *World Health Report 1999*. Geneva.

WHO (World Health Organization). 1999b, June 17. "Infectious Diseases: WHO Calls for Action on Microbes." Geneva.

WHO (World Health Organization) 1999c. "Meningococcal and Pneumococcal Information Page." At http://www.who.int/gpv-dvacc/research/mening.html.

WHO (World Health Organization). 1997a. *Weekly Epidemiological Report* 72(36–8).

WHO (World Health Organization). 1997b. *Anti-Tuberculosis Drug Resistance in the World*. Geneva.

WHO (World Health Organization). 1996a. *Investing in Health Research and Development: Report of the Ad Hoc Committee on Health Research Relating to Future Intervention Options*. Geneva.

WHO (World Health Organization). 1996b. *World Health Organization Fact Sheet N94 (revised)*. Malaria. Geneva.

WHO (World Health Organization) and UNICEF. 1996. *State of the World's Vaccines and Immunization*. WHO/GPV/96.04. Downloadable from http://www.who.int/gpv-documents/docspf/www9532.pdf.

World Bank. 1993a. *Disease Control Priorities in Developing Countries*. Oxford Medical Publications. New York: Published for the World Bank by Oxford University Press.

World Bank. 1993b. *World Development Report 1993: Investing in Health*. Washington, DC: Oxford University Press.

World Bank. 1998. *World Development Indicators*.

World Bank. 1999. "Preliminary Ideas on Mechanisms to Accelerate the Development of an HIV/AIDS Vaccine for Developing Countries."

World Bank AIDS Vaccine Task Force. 2000, February 28. "Accelerating an AIDS Vaccine for Developing Countries: Recommendations for the World Bank."

Wright, Brian D. 1983, September. "The Economics of Invention Incentives: Patents, Prizes, and Research Contracts." *American Economic Review* 73:691–707.

3

Creating Markets for New Vaccines
Part II: Design Issues

Michael Kremer, *Harvard University, The Brookings Institution, and NBER*

Executive Summary

Several programs have been proposed to improve incentives for research on vaccines for malaria, tuberculosis, and HIV, and to help increase accessibility of vaccines once they are developed. The U.S. administration's 2000 budget proposed a tax credit that would match each dollar of vaccine sales with a dollar of tax credit. The President of the World Bank has proposed a $1 billion fund to provide concessional loans to countries to purchase vaccines if and when they are developed. European political leaders have spoken favorably about the concept of a vaccine purchase fund. This paper explores the design of such programs, focusing on commitments to purchase new vaccines.

For vaccine purchase commitments to spur research, potential vaccine developers must believe that the sponsor will not renege on the commitment once vaccines have been developed and research costs sunk. Courts have ruled that similar commitments are legally binding contracts. Given appropriate legal language, the key determinant of credibility will therefore be eligibility and pricing rules, rather than whether funds are physically set aside in separate accounts. The credibility of purchase commitments can be enhanced by specifying rules governing eligibility and pricing of vaccines in advance and insulating those interpreting these rules from political pressure through long terms.

Requiring candidate vaccines to meet basic technical requirements, normally including approval by some regulatory agency, such as the U.S. FDA, would help ensure that funds were spent only on effective vaccines. Requiring developing countries to contribute copayments would help ensure that they felt that the vaccines were useful given the conditions in their countries.

The U.S. Orphan Drug Act's success in stimulating research and development is widely attributed to a provision awarding market exclusivity to the developer of the first drug for a condition unless subsequent drugs are clinically superior. Purchases under a vaccine purchase program could be governed by a similar market exclusivity provision.

A purchase commitment program could start by offering a fairly modest price. If this proved inadequate to spur sufficient research, the promised price could be increased. This procedure mimics auctions, which are often efficient procurement methods when costs are unknown. As long as prices do not rise at a rate substantially greater than the interest rate, vaccine developers would not have incentives to withhold vaccines from the market.

The World Bank has termed health interventions costing less than $100 per year of life saved as highly cost-effective for poor countries. If donors pledge approximately $250 million per year for each vaccine for 10 years, vaccine purchases would cost approximately $10 per year of life saved. It is unlikely that vaccines for all three diseases would be developed simultaneously, but if donors wanted to limit their exposure, they could cap their total promised vaccine spending under the program, for example at $520 million annually. No funds would be spent or pledges called unless a vaccine were developed.

I. Introduction

Several initiatives have recently been proposed to create incentives for research on vaccines against diseases such as malaria, tuberculosis, and AIDS, and to increase accessibility of vaccines once they are developed. The president of the World Bank recently said that the institution is planning to establish a $1 billion fund to help finance purchases of new vaccines, if and when they are developed (Financial Times 2000) although the Bank has not yet acted on this initiative. The U.S. administration's 2000 budget included a tax credit for vaccine sales that would effectively double the developing country market for new vaccines against diseases that kill more than one million people each year (http://www.treas.gov/taxpolicy/library/grnbk00.pdf). The tax credits would be capped at $1 billion over 10 years. The concept of a vaccine purchase fund has also received support from European political leaders (www.auswartiges.amt.de, 1999; DFID 2000).

Although malaria, tuberculosis, and African strains of AIDS kill almost 5 million people each year, they are the subject of little vaccine research. Potential vaccine developers fear that they would not be able to sell enough vaccines at a high enough price to recoup their research investments. This is both because these diseases primarily affect poor countries, and because vaccine markets are severely distorted. The proposed programs could both create incentives for vaccine research and help improve access to any vaccines developed (see the companion paper, "Creating Markets for New Vaccines: Part I: Rationale"). They would not require any expenditure unless and until vaccines were developed.

This paper addresses the many design issues that would arise in establishing such programs. It focuses on the design of a vaccine purchase commitment, but much of the analysis carries over to the analysis of tax credits and a World Bank loan fund. Policymakers considering establishing such programs face a host of questions. How can commitments be made credible to vaccine developers? How should eligibility of candidate vaccines be determined? What prices should be paid for vaccines, and should these prices vary with vaccine characteristics? If multiple vaccines are developed, which should be purchased? Should recipient countries provide copayments, and if so, how much? How cost-effective would such programs be?

If potential vaccine developers are to invest in research, they must believe that once they have sunk funds into developing a vaccine, the sponsors of a vaccine purchase program will not renege on their commitments by paying a price that covers only the cost of manufacturing, and not research. Section II of this paper discusses factors affecting the credibility of a vaccine purchase commitment. Courts have held that similar public commitments to reward contest winners or to purchase specified goods constitute legally binding contracts and that the decisions of independent parties appointed in advance to adjudicate such programs are binding. This suggests that if programs contain appropriate legal language, the key determinant of their credibility will not be whether funds are physically set aside in a separate account, but the rules determining eligibility and pricing, and the procedures for adjudicating decisions under these rules. If potential vaccine developers are to invest in research, they must be confident that the adjudicators will not abuse their power. The credibility of a vaccine purchase commitment can be enhanced by clearly specifying eligibility and pricing rules, insulating decision makers from political pressure through long terms of service, and including former industry officials on the adjudication committee.

Section III argues that requiring countries that receive vaccines to provide copayments in exchange for vaccines will give countries incentives to carefully investigate whether candidate vaccines are appropriate for their local conditions. Moreover, for any fixed level of donor contributions, requiring copayments gives potential vaccine developers greater incentives to conduct research.

Section IV outlines a possible process for determining vaccine eligibility and pricing. Candidate vaccines would first have to meet some minimal technical requirements, which would ordinarily include clearance by a regulatory agency, such as the U.S. Food and Drug Administration (FDA). They would then be subject to a market test: Nations wishing to purchase vaccines would need to provide a modest copayment tied to their per capita income and spend down an account assigned to them within the program. Any vaccine meeting these requirements would be eligible for purchase at some base price. Vaccines exceeding these minimum requirements could potentially receive bonus payments tied to vaccine effectiveness.

Section V discusses procedures if multiple vaccines are developed for a single disease. The developer of the first vaccine against a disease creates enormous social benefits. Developers of subsequent vaccines create benefits only to the extent that their vaccines are superior or serve populations that are not served by the first vaccine. This suggests that rewards should be greatest for the first vaccine developer. The U.S. Orphan Drug Act specifies that the first developer has market exclusivity unless a subsequent product is clinically superior. This provision is generally believed to account for the Act's success in increasing research on orphan drugs. An analogous provision could grant market exclusivity for purchases under the program to the first vaccine developed unless subsequent vaccines were clinically superior.

Section VI discusses vaccine pricing and coverage. Research and development on vaccines is typically very expensive, but manufacturing additional doses is usually reasonably cheap. Given total revenue from a vaccine, research incentives are likely to be fairly similar if few doses are sold at a high price, or many doses are sold at a lower price. This suggests that it is efficient to pay per immunized child, rather than per dose, and to include countries and demographic groups in the program as long as vaccination is cost-effective at the incremental cost of producing additional doses (rather than at the average price per person immunized paid under the program). The total market promised by the program should be large enough to induce substantial effort by vaccine developers, but less than the social value of the vaccine. A rough rule of thumb in the industry is that a $250 million annual market is needed to motivate substantial research. A program in which donors provide approximately $250 million in average annual contributions and copayments average another $86 million annually would cost approximately $10 per year of life saved. The World Bank has termed health interventions costing less than $100 per year of life saved as highly cost-effective (World Bank 1993).

One way to avoid either paying more than necessary for a vaccine or offering too little to stimulate research would be to offer a relatively modest price initially, and if this price proved insufficient, to raise the promised price gradually until it proved sufficient to spur vaccine development. As long as the price did not increase at a rate substantially greater than the interest rate, vaccine developers would not have incentives to withhold vaccines from the market in hopes of obtaining a higher price. It is unlikely that vaccines for all three diseases would be developed simultaneously, but if donors wished to limit their potential liability, they could cap their committed annual expenditure.

Section VII discusses the appropriate scope of vaccine purchase commitments. Should the program be limited to vaccines, or also include drugs? Which diseases should be covered?

The conclusion briefly considers the politics of programs to improve vaccine markets. It then discusses the proposed U.S. and World Bank programs and how a private foundation could participate in a purchase commitment program.

This paper builds on earlier work. The idea of an HIV vaccine purchase program was discussed in WHO 1996, and advocated by a coalition of organizations coordinated by the International AIDS Vaccine Initiative (IAVI) at the 1997 Denver G8 summit. Since then, the idea has been explored by the World Bank AIDS Vaccine Task Force (World Bank 1999, 2000). Kremer and Sachs (1999) and Sachs (1999) have discussed such programs in the popular press. This paper also draws on earlier work on vaccines, including Batson 1998, Dupuy and Freidel 1990, Mercer Management Consulting 1998, and Milstien and Batson 1994, and on the broader academic literature on research incentives, including Guell and Fischbaum 1995, Johnston and Zeckhauser 1991, Lanjouw and Cockburn 1999, Lichtmann 1997, Russell 1998, Scotchmer 1999, Shavell and van Ypserle 1998, and Wright 1983.

II. The Credibility of Vaccine Purchase Commitments

For a vaccine purchase commitment to be effective in spurring new research, potential vaccine developers must believe that once they have sunk money into producing a vaccine, it will be purchased at a price that covers their risk-adjusted costs of research, as well as their manufacturing costs. The first subsection, titled Legal Doctrine, notes that courts have held similar commitments to be legally binding contracts and argues that as long as the sponsor of a commitment has sufficient funds to fulfill the commitment, physically moving money to a separate account is unnecessary to provide legal commitment. The second subsection, titled Issues to Consider in Determining Eligibility and Pricing, discusses some of the issues that would need to be addressed in specifying eligibility and pricing rules based on technical characteristics of a vaccine. The third subsection, titled Procedures to Increase Credibility of a Vaccine Purchase Commitment, argues that some discretion will be needed to interpret how general eligibility and pricing rules apply to any specific candidate vaccine, and discusses how the credibility of adjudicating institutions could be enhanced.

Legal Doctrine

This section argues that a suitably designed commitment will be interpreted by the courts as a legally binding contract, and that hence the key credibility issue will not be outright default by the program sponsor, or whether money is physically set aside in a separate vaccine purchase fund, but rather questions over the interpretation of program rules.

Courts have ruled that publicly advertised contests are legally binding contracts (Morantz and Sloane 2000). As summarized in Sullivan 1988, sponsors of contests are contractually obligated to pay the winners according to their public announcements. A contestant who performs the requested act has formed a valid and binding contract with the promoter. Attempts to escape liability by changing contest rules after a contestant has accepted the offer by performing the desired act are generally treated as breach of contract. Advertisements with certain specifications (identification of good, definite quantity of good, etc.) for the purchase of goods at specified prices have also been found to be legally binding. (See Vaccaro 1972 for a summary and analysis of doctrine.)

Moreover, if the procedures in a contest stipulate who will judge the contest, decisions made by the stipulated judge of the contest are usually treated as conclusive (Morantz and Sloane 2000). The majority view among courts is that judges' decisions are conclusive as long as they are made in good faith, although some cases find that contracts giving one party the unilateral right to decide disputes are unenforceable. When the judge of the contest is an independent party, the courts almost universally hold the decision as final unless the decision was made in bad faith, or the judges exceeded the authority specified in contest rules.[1]

There are a number of precedents for programs to reward developers of new technologies. The British government established a £20,000 prize for a method of determining longitude at sea after a fleet got lost and struck rocks, drowning 2,000 sailors. The prize was won by John Harrison for the chronometer.[2] More recently, the Kremer prize for human-powered flight led to the historic flight of the Gossamer Albatross across the English Channel (Grosser 1991). The $30 million "golden carrot" tournament for an energy efficient refrigerator sponsored by 24 U.S. electric utilities offered to pay the winning team a certain amount for every unit sold. Whirlpool won the tournament with a line of refrigerators that operated 70% more efficiently than 1992 federal requirements.

Given that legally binding contracts can be written, physically setting aside funds in an escrow account is not necessary for a commitment to be binding, as long as the sponsor of a vaccine purchase commitment has sufficient funds to fulfill the commitment. The key questions for credibility revolve around specifying eligibility and pricing rules and procedures for adjudicating disputes under these rules.

Depending on legal language, commitments could be made more or less binding. The options range from a simple announcement of an intention to purchase vaccines, to a legally binding announcement with details on eligibility and pricing. The more binding the commitment, the stronger the incentives for potential vaccine developers. In general, there is a trade-off between flexibility and credibly committing to pay for a vaccine. Imperfect commitment reduces both the expected revenue for vaccine developers and expected costs for the sponsor in the same proportion. It reduces efficiency to the extent that the parties are risk averse.[3]

Issues to Consider in Determining Eligibility and Pricing

A program to increase the market for vaccines could offer to purchase vaccines meeting certain technical specifications, offer to match money spent on vaccine purchases by other institutions, or use some combination of these approaches. For example, the Kremer prize laid out detailed technical eligibility requirements. The U.S. tax credit proposal does not specify detailed technical requirements, other than FDA approval, but merely states that a 100% tax credit will be given for sales of vaccines to nonprofits and international institutions, which would presumably make their own judgments as to whether candidate vaccines are acceptable.

The following are some of the key issues which would need to be considered in determining vaccine eligibility and pricing based on technical specifications.

- vaccine efficacy—the reduction in disease incidence among those receiving the vaccine.[4] Efficacy might vary in different circumstances. A vaccine could potentially be more efficacious against some strains of the disease than others, and thus be better suited to some geographic areas than others. It could work for some age groups, but not others. A vaccine might prevent severe symptoms of the disease, but not prevent milder cases.

- the number of doses required, the efficacy of the vaccine if an incomplete course is given, and the ages at which doses must be taken. If too many doses are required, fewer people will bring their children in to receive the full course of immunization. If the vaccine can be given along with vaccines that are already widely administered, delivery will be much cheaper.

- vaccine side effects. Side effects could differ for different subpopulations. Side effects would also need to be considered for people who do not comply perfectly with the delivery protocol. For example, taking a partial course of a malaria vaccine could potentially interfere with natural limited immunity.

- the time over which the vaccine provides protection, and whether booster shots could extend this period.

- what level of rigor would be required in the field trials. For example, how long would subjects have to be followed to determine the length of protection? How many separate studies in different regions would need to be conducted to assess efficacy against different varieties of the disease?

- the extent to which vaccines would lose their effectiveness over time. Presumably, some ongoing monitoring of vaccine effectiveness in the field would be required, and if it appears that resistance to the vaccine is spreading, vaccine purchases would have to be reassessed.

One possibility would be to design eligibility rules using these criteria in such a way that vaccines would be considered eligible if they met a cost-effectiveness threshold.[5] Eligibility and pricing rules could potentially be set so that vaccines meeting a certain cost-effectiveness threshold would be eligible for purchase and vaccines exceeding this threshold would receive higher prices.

Note, however, that misspecifying eligibility and pricing rules could misdirect research incentives away from appropriate vaccines, or vitiate research incentives altogether. For example, if the program failed to specify otherwise, it might be obligated to purchase a malaria vaccine that interfered with the development of natural immunity and provided only temporary protection. Such a vaccine might merely postpone malaria deaths. If such a vaccine were eligible for purchase under the program, researchers might pursue it, rather than devoting their efforts to more useful lines of research. On the other hand, there is a risk that the program could set specifications so stringent that they would be difficult to achieve. This would discourage pharmaceutical firms from following promising leads. For example, if the specifications required 90% efficacy against all strains of the disease, potential vaccine developers might not pursue a candidate vaccine that would be likely to yield 99% protection against most strains, but only 85% protection against others. If it were difficult to create a vaccine delivering 90% protection in all regions, no vaccine at all might be developed.

Aside from specifying eligibility rules, the program would have to specify pricing rules. Paying more for superior vaccines might create more appropriate incentives for researchers. A 90% efficacious vaccine is worth more than an 80% efficacious vaccine, and a vaccine that requires no booster is worth more than one requiring boosters every 5 years.

Procedures to Increase Credibility of a Vaccine Purchase Commitment

General eligibility and pricing rules could be set out, but someone would have to exercise discretion in interpreting these rules once vaccines have been developed and tested.[6] Once the vaccine developer has sunk hundreds of millions of dollars in research, adjudicators might be tempted to offer a price that covered only manufacturing costs or to insist on excessive product testing and improvements. If pharmaceutical executives suspect that the adjudicators will succumb to these temptations, the companies will be reluctant to invest in a vaccine.

Credibility of vaccine purchase commitments to potential developers could be enhanced by appointing appropriate decision makers (such as a committee with some members who have worked in the pharmaceutical industry), insulating decision makers from political pressures through long terms of service, establishing a minimum purchase price, and placing limits on the discretion of the program committee by laying out reasonably transparent rules for determining eligibility and pricing. Another way to enhance the credibility of a commitment is to establish a program that covers a number of different diseases which primarily affect developing countries. The program would then have an incentive to build up a reputation for fair play.[7]

The experience of central banks may offer some lessons for the design of a vaccine purchase program. Just as a vaccine purchase program would need to make a credible commitment to purchase an effective vaccine if one were developed, central banks need to head off inflationary expectations by credibly promising to take tough action if inflation starts to increase. Central banks insulate decision makers from political pressures by appointing them for long terms, and a vaccine purchase program could do the same. Appointing central bankers with strong anti-inflation credentials also helps build credibility for central banks. Similarly, delegating decisions regarding eligibility and pricing to a committee which included some members who had worked in industry might help convince potential vaccine developers that the committee would not impose unreasonable conditions after they developed a vaccine.[8]

Commitments by the vaccine purchase program will be more credible if the program administrators have incentives to build a reputation for fulfilling promises. If the program covered vaccines against several diseases, program administrators would have incentives to develop a reputation for treating vaccine developers fairly, so as to build credibility with potential developers of other vaccines.[9] Nonetheless, it may take time to develop a reputation.

One way to help build credibility with potential vaccine developers would be to set a minimum price in advance.[10] This could help solve the time-consistency problem, but at some cost. A vaccine which is useful, but not useful enough to warrant purchase at the minimum guaranteed price, would not be purchased at all. In practice, however, this problem may not be that serious. Most vaccines that passed regulatory approval would be cost-effective at even a high price per person immunized relative to the likely availability of funds (Glennerster and Kremer 2000). This is because vaccines falling far short of U.S. or European regulatory requirements have great difficulty winning wide approval in developing countries in any case.[11] If one takes as given that vaccines will only be used if they meet a stringent risk-benefit ratio, it seems quite unlikely that guaranteeing a minimum price *ex ante* would lead to rejection of an otherwise usable vaccine on cost-effectiveness grounds. If a vaccine were not useful enough to warrant purchase at a few dollars per person immunized, the cost of failing to purchase it would not be that great. Moreover, if a vaccine turned out to be socially useful, but not good enough to qualify for purchase under the program at the promised price, this would not preclude individual countries from purchasing the vaccine or other donors from purchasing it. The costs of guaranteeing a minimum price seem small relative to the benefit of improving the credibility of commitments to reward vaccine developers, and thus spurring research.

III. Copayments

Another way to increase the market for vaccines would be to offer to match others' expenditures on vaccine purchases. This is similar, in effect, to purchasing the vaccine and providing it to others in exchange for a copayment. Requiring countries receiving vaccines to provide reasonable copayments can boost incentives for vaccine developers given any fixed level of donor contributions. Copayments also help ensure that the authorities in recipient countries feel that the vaccine is suitable for use in their circumstances. This is important since conditions vary among countries. For example, a vaccine might be effective against the strains of malaria prevalent in some countries, but not against other strains. Finally, requiring copayments is a useful test of a country's commitment to a program. If a country is prepared to make a copayment, it is also more likely to be prepared to take the other steps necessary to ensure that the vaccine is delivered to the people who need it.

Setting the level of copayments involves a trade-off between improving access once a vaccine has been developed and creating incentives for vaccine development. On the one hand, once a vaccine has been developed, it will be produced at the efficient scale if the copayment equals the marginal cost of producing an additional dose. On the other hand, given a fixed level of donor contributions, incentives for vaccine development will be greater if developing countries provide copayments at their willingness to pay for the vaccine.

Setting copayments from countries receiving vaccines just below their estimated willingness to pay for vaccines will maximize incentives for vaccine development while not reducing consumption of vaccines below the optimal level. Since richer countries are likely to be willing to pay more for vaccines than poorer countries, this implies that copayments should rise with per capita income.[12] Willingness to pay may also be greater for diseases that create a particularly high health burden, such as HIV/AIDS. Given the uncertainty in estimating this willingness to pay and the need for a uniform copayment policy across heterogeneous countries, it makes sense to estimate willingness to pay conservatively. Insisting on too great a copayment would limit access to the vaccine, and, by reducing take-up, would reduce incentives for vaccine developers.

Note also that setting the required copayments close to countries' willingness to pay reduces vaccine developers' temptation to try to extract supplemental payments from purchasing countries. It is not clear whether the vaccine purchase program should agree to be a party to vaccine purchases with supplemental copayments greater than those required under the program, even if the recipient country agrees to this. Allowing supplemental payments broadens the scope for vaccine developers to demand prices greater than those offered under the vaccine purchase program, and these higher prices could potentially exclude some countries from access to vaccines. For example, if the vaccine developer felt that most countries would be willing to supplement the required copayment by $1 a dose, it might demand this from every country. Those countries unable to afford this supplemental payment would not be able to obtain the vaccine.

Note that tying copayments to income achieves many of the benefits of tiered pricing. If copayments are set appropriately, access to vaccines is expanded so that vaccines can be used wherever the social value of the vaccine exceeds the marginal production cost. Incentives for vaccine development can correspond to the aggregate willingness to pay for vaccines. Yet vaccine developers need not take the politically damaging step of revealing their willingness to produce additional doses at low cost, thus risking generating enhanced political pressures for price regulation.

IV.   Combining Technical Requirements and a Market Test

Technical eligibility requirements could potentially be combined with a market test. For example, candidate vaccines could first be required to meet basic technical requirements, which would typically include clearance by some regulatory agency (such as the U.S. FDA). They could then be required to meet a market test—developing countries wishing to purchase vaccines using program resources would be required to contribute a copayment, and would be required to draw down an account they would have within the vaccine purchase program. Any vaccines meeting these requirements would be eligible for purchase at some base price. Vaccines exceeding these requirements could potentially receive bonus payments linked to vaccine effectiveness. Ideally, this would make commitments to purchase useful vaccines at remunerative prices credible to potential vaccine developers, but would leave enough flexibility that appropriate purchasing decisions could be made after vaccines had been tested and their characteristics became known.

Basic Technical Requirements

To be eligible for purchase, vaccines could be required to fulfill basic technical requirements, which would normally include regulatory clearance by an established regulatory agency, such as the U.S. FDA or its European counterpart. This would ensure that the funds were spent for *bona fide* vaccines, rather than for quack remedies. However, a vaccine may pass a risk-benefit assessment in one country, but not another. For example, a malaria or tuberculosis vaccine with significant but small side effects might not be appropriate for general use in low prevalence countries, such as the United States, but might save millions of lives in high prevalence areas.

It might make sense to allow the program, at its discretion, to waive the requirement of regulatory approval in donor countries if a country requested the vaccine and a scientific committee established by the program concurred that the vaccine satisfied the risk-benefit assessment given the situation in the applicant country. More generally, it might be appropriate to guarantee that any candidate vaccine satisfying certain high technical standards would receive automatic approval to go on to the market test. There could also be a gray area in which candidate vaccines could be approved at the discretion of a scientific committee. This would provide assurance to potential vaccine developers that if they develop a high-quality vaccine, they will have a market. It would also allow the committee the flexibility to consider purchasing vaccines that passed a risk-benefit analysis, but fell short of an ideal vaccine.

Just as a vaccine might satisfy a risk-benefit assessment in a high prevalence country, but not in the United States, it is possible that a vaccine could be appropriate in the United States, but not elsewhere. For example, a malaria vaccine that interfered with natural immunity might be appropriate for U.S. travelers, who would not have built up this immunity in any case, but not for long-term residents of malarious areas. A few minimal technical requirements beyond regulatory approval are therefore likely to be appropriate before vaccines were made eligible for the market test described below. Travelers' vaccines for malaria, which protect people making short trips, would presumably be ineligible.[13] Other technical requirements might include a requirement that a vaccine could only be purchased for a country if it had been shown effective for the strains of disease prevalent in that country. Vaccines requiring more than some cutoff number of doses to be effective might require a special waiver for eligibility. Some ongoing monitoring might be required, to ensure that resistance to the vaccine had not developed and spread. However, the credibility of purchase commitments with potential vaccine developers would be enhanced by keeping technical eligibility requirements beyond regulatory clearance minimal and clearly defined to reduce the potential for abuse of discretion. This would decentralize the basic purchasing decision to individual recipient countries. Of course, these countries would be free to consider recommendations put out by the World Health Organization or any other body.

The Market Test

As discussed above, vaccines could meet regulatory approval, but still be unsuitable for widespread use in a particular developing country. For example, a vaccine that was effective only if people received ten precisely timed doses might be useful for the U.S. military, but not for most people in developing countries. Requiring vaccines which satisfy the technical criteria to meet a market test would allow purchasers the flexibility to make decisions about whether a particular vaccine is appropriate for their needs. In particular, developing countries would have incentives to seriously consider the suitability of candidate vaccines if they had to provide a copayment and draw down an account within the vaccine purchase program that would be established specifically for each country.

Copayments help ensure that after a particular vaccine has been tested, it is considered worth purchasing. However, copayments alone may not be sufficient to demonstrate country commitment, since donors might offer to help fund copayments. It is not clear that it would be possible or desirable to prohibit this.

Countries could be further motivated to carefully consider their purchases by establishing subaccounts within the program for each country. If a country decided to purchase a vaccine, it would draw down the commitments allocated to it. This system would give countries an incentive to purchase a vaccine only if they were confident that it could be effectively administered in their country and if they did not expect a superior vaccine to come on the market shortly. Otherwise, they would be better off saving the funds in their subaccount.

In the absence of separate subaccounts, countries might agree to purchase even marginally effective vaccines, knowing that if they did not consume the available funds, other countries would. If potential vaccine developers anticipated this, they might invest in a candidate vaccine that looked like it would meet only minimal eligibility requirements, rather than investing in a slightly more risky, but ultimately much more promising, vaccine. If countries must spend funds earmarked for their own vaccine purchases, they will have more incentive to purchase only high quality vaccines, thus providing incentives for potential vaccine developers to focus on developing such vaccines. Since countries would not be able to use their accounts to purchase anything but vaccines, and would not receive interest on their accounts if they remained unspent, they would have every incentive to use their accounts to purchase a good vaccine if one were available.[14]

Relying only on a market test and eliminating any technical requirements could potentially lead to the purchase of inappropriate vaccines due to bribery or tied deals. Vaccine developers could offer to kick back some percentage of the purchase price to the developing country in the form of price reductions on other pharmaceuticals, or even bribes. This could potentially be an attractive arrangement for the developing country or its officials, since the country itself would contribute only a copayment toward the cost of the vaccine, with the bulk of the financing coming from the vaccine purchase program.

A series of safeguards are therefore needed to prevent purchase of inappropriate vaccines due to bribery or tied deals. The technical requirements for eligibility provide the first and most important line of defense. This would prevent a country from using program funds to purchase a quack vaccine manufactured by a politically connected firm. Outright corruption could probably be limited with provisions punishing firms found guilty of bribing officials and restricting the amount of travel, training, and other perks that vaccine sellers could provide to health ministry officials. Under the U.S. Foreign Corrupt Practices Act, firms and executives found guilty of bribing foreign governments are subject to criminal prosecution. Other nations are now adopting similar laws. Since the developing country vaccine market is a small part of overall business for most large pharmaceutical companies, they would likely be reluctant to risk bad publicity, the attention of regulators, and legal sanction in order to make some extra money on vaccines.

Whistle blower procedures could be instituted to protect, or even reward, committee members reporting attempts at bribery by vaccine developers. Similarly, vaccine developers could blow the whistle on committee members who tried to insist on kickbacks. Members of the committee who were proven to have asked for kickbacks could be removed from the committees.

Implicit tied dealings are more difficult to regulate. A pharmaceutical firm simultaneously negotiating with a health ministry over a malaria vaccine and an antibiotic might convey to the ministry that it would be willing to be flexible on the antibiotic price if the ministry would purchase the malaria vaccine. In the absence of further incentives, vaccine developers might therefore aim only at creating a vaccine that could pass minimal eligibility requirements, rather than a more widely useful vaccine.[15]

One way to limit corruption and tied deals, while still preserving a market test, would be to include civil society as well as governments in countries' decision making processes. For example, the committee making purchase decisions for a country might include not only representatives of the Ministry of Health, but also respected physicians, nongovernmental organization representatives, and scientists. Countries wishing to participate in the program could be required to set up such committees in advance, and members could have security of tenure. Some members of the committee could be appointed by the vaccine purchase program. The committee could have authority to release resources from the country's subaccount within the program. The government would need to authorize disbursements of public funds to cover copayments, but donors could potentially fund copayments.

Limiting the number of doses purchased for any one country would limit the potential loss from tied deals and corruption. The number of doses purchased for a country might be limited to the number needed for the annual birth cohort, with some adjustment for the initial years of the program when a backlog of unimmunized people would need to be vaccinated.

Bonus Payments Based on Vaccine Quality

Specifying a minimum price which would be paid for vaccines meeting the first two steps—technical requirements and the market test —would help provide potential developers with a credible commitment. However, it would be desirable for developers to have incentives to develop vaccines that exceed a minimum eligibility threshold. It might therefore be useful to provide bonus payments depending on vaccine quality. One standard way to measure cost-effectiveness in health is the cost of saving a Disability Adjusted Life Year, or DALY. DALYs take into account not only the years of life lost but also the years of disability caused by a disease. In order to create appropriate incentives for vaccine developers to develop high quality vaccines, bonus payments could be set so as to tie the reward to the number of lives or DALYs saved and to the cost of delivery.

Bonuses could be provided for vaccines believed to exceed a cost-effectiveness threshold, in dollars spent per DALY averted. If a vaccine exceeded this threshold, some fraction of the resulting savings could be returned to the vaccine developer as a bonus above the base price.

Basing incentives on lives or DALYs saved would create good incentives for pharmaceutical firms to develop vaccines that create positive externalities, such as a malaria vaccine with an altruistic component which kills gametocytes, and thus prevents other people from becoming infected. Any side effects of a vaccine could be subtracted from the measure of lives or DALYs saved.[16]

Bonuses could also be paid if the vaccine were cheap to deliver. This would create incentives for researchers to develop vaccines that are oral, rather than injectable, that do not require many doses, and that can be delivered along with the vaccines currently given as part of the Expanded Program on Immunization (EPI).

Bonus payments could potentially be set in two ways. A committee could be free to base bonus payments directly on its estimates of the number of lives or DALYs saved by a particular vaccine, using any data it wished.[17] Alternatively, a schedule of bonus payments could be set in advance as a function of more easily measured vaccine characteristics, such as efficacy in clinical trials, the number of doses needed, etc. An approach such as that used by Glennerster and Kremer (2000) could be extended to estimate the set of vaccine characteristics associated with any particular cost-effectiveness threshold.

Directly estimating DALYs or lives saved after vaccines are developed allows the program to consider a broad range of vaccine characteristics and to use up-to-date information, but it also creates more uncertainty for vaccine developers and raises the prospect of bias by the committee charged with estimating DALYs and costs.[18] The appropriate strategy depends in part on how trustworthy the committee charged with these tasks is considered to be, and in part on what reasonably transparent and objective procedures can be developed for measuring vaccine efficacy. Thus, it may vary among diseases.[19]

Basing payments directly on the number of lives or DALYs saved through a vaccine and the cost of delivery is also potentially problematic because these quantities depend not only on actions under the control of the vaccine developer, but also on actions by others. To the extent that health ministries cannot easily maintain cold chains or deliver vaccines to rural areas on a precise schedule, vaccinations that require cold chains and precisely timed deliveries will be expensive per life or DALY saved.

If the weaknesses of health ministries are not strategically aimed at extracting payments from the vaccine developer, this will create appropriate incentives for vaccine developers. Vaccine developers should try to design vaccines that are appropriate for actual health systems, not for some theoretical ideal health system. For example, if health ministries cannot maintain cold chains for vaccines, then vaccine developers should have incentives to develop heat stable vaccines.

However, to the extent that health ministries behave strategically, it will be best to base bonus payments on preset indicators of the likely number of DALYs saved, rather than the actual number of DALYs. This is because if vaccine developers were paid based on realized DALYs saved, health ministries could potentially try to extract payments from the vaccine developer in exchange for agreeing to distribute the vaccine efficiently. This would weaken incentives for vaccine development.

If the committee charged with estimating lives or DALYs saved simply makes honest mistakes in calculating these quantities, but those mistakes do not systematically tend to underestimate or overestimate the actual effects of the vaccine, then the potential profit from developing a vaccine could as easily be increased or decreased by the uncertainty in calculations of DALYs or lives saved. The attractiveness of investment in vaccines would be reduced, but only to the extent that vaccine developers are not willing to take gambles that could turn out to help them as easily as to hurt them.

Errors in estimation of DALYs or lives saved are particularly problematic if vaccine developers can influence these estimates through actions other than research. For example, if politically connected pharmaceutical firms obtain more favorable DALY calculations, firms will divert effort towards developing political connections and away from developing good vaccines.

The scope for bias would be reduced by setting forth procedures as fully as possible ahead of time, working under a framework of establishing a bonus per life or DALY saved. The World Health Organization project on the burden of disease has developed detailed procedures for estimating DALY burdens. Epidemiological surveys could be conducted to assess the burden of various diseases prior to the development of vaccines.

Sunset Provisions

Sunset provisions could be incorporated into a vaccine purchase program. For example, a malaria vaccine fund could revert to the donors or be used for other health problems in developing countries if after 50 years no qualifying vaccine had been developed, or if at some earlier time a scientific committee established by the program determined that the burden of malaria had been sustainably cut more than 50% through other techniques, such as insecticides. Sunset provisions could be continuous, so that the purchase commitment would fall with the severity of the disease. Note that any bonus payment based on DALYs or lives saved would automatically fall with prevalence of the disease. A sunset provision would increase the risk borne by potential vaccine developers, but biotech and pharmaceutical firms routinely have to bear the risk that alternative technologies will render the projects they are working on superfluous. There is no reason why this should be any different for firms working on developing country diseases. It is efficient for researchers to consider the possibility that their work will be superseded by other technologies when choosing their research projects.

V. Procedures for Multiple Vaccines

For vaccine purchase commitments to spur research, it is essential that intellectual property rights be respected. If the program purchases vaccines from imitators, rather than respecting the intellectual property rights of the original developers, incentives for vaccine development will be vitiated.[20]

However, enforcing patents may not be enough. Once one vaccine for a disease has been developed, it becomes easier for competitors to develop alternative vaccines, even if the first is protected by a patent, as it can be relatively easy to design around vaccine patents. Developers of the initial vaccine, therefore, face a risk that a marginally superior vaccine will be produced shortly after the initial vaccine is developed and that this subsequent vaccine will capture the entire market. This risk may deter research. In many industries, first mover advantages due to network effects or to brand loyalty by customers are as important as patents in protecting innovations, but since governments are the main purchasers of vaccines, and are less likely to be influenced by brand loyalty, other forms of protection may be needed for vaccine developers.

It will be important to preserve rewards for the initial developer, who will have made the largest investment in research. Currently, the world needs acceptable vaccines for malaria, tuberculosis, and HIV/AIDS, and incentives for a private developer are a small fraction of the social value. Once an adequate vaccine is developed, however, the world's need for a second vaccine will be much more limited. This suggests a smaller reward will be needed to bring private incentives for a second vaccine into line with the social value of a second vaccine. To some extent, the initial developer will receive a larger share of vaccine purchases in any case, since the initial developer will sell vaccines used to immunize the backlog of unimmunized adults, while subsequent developers will be restricted to the market of new cohorts of children. Pricing a vaccine in nominal terms will also disproportionately help the original developer, since real prices will fall over time. (It would even be possible to specify a falling time path of nominal prices.)

The developer of the first vaccine could be further protected through an exclusivity clause similar to that in the Orphan Drug Act. This would require that the initial vaccine be purchased if newer alternatives were not clinically superior. This provision is widely believed to have greatly increased research on orphan drugs (Shulman and Manocchia 1997).

In practice, the exception for "clinically superior" vaccines may not weaken incentives for the first developer that much, since regulatory standards for approval of the first vaccine are likely to be high, and it may be difficult to show that a subsequent vaccine is "clinically superior."

Note that market exclusivity would apply only to the target population for which the original vaccine was adequate. Thus, for example, if one firm develops an AIDS vaccine effective against a particular clade of the disease, it would have marketing exclusivity for that clade, but not for other clades.

One potential objection to the market exclusivity provision is that it could increase the risk borne by developers. In the absence of a market exclusivity clause, if several firms develop vaccines around the same time, they will share the market. Providing market exclusivity to the first vaccine developer could potentially increase risk. On the other hand, to the extent that prices fall if multiple vaccines are invented, or firms dissipate potential profits in marketing expenditures, the expected reward to investing in vaccine research and development is greater with a market exclusivity clause. The success of the Orphan Drug Act in increasing research and development on orphan drugs suggests that the increase in expected profits is the key issue for potential developers. If it were thought important to avoid increasing the risk borne by potential vaccine developers, purchases under the program could be limited to those vaccines invented within some period (perhaps a year or two) following the licensing of the first acceptable vaccine, unless a subsequent vaccine was clinically superior. This would reduce risk for firms engaged in a tight race to develop a vaccine, while also reducing the chance that "me too" vaccines would greatly reduce sales for the initial developer and thus deter research.

The exception in the Orphan Drug Act's market exclusivity provision for clinically superior products could potentially be modified for application to a vaccine purchase commitment. Ideally, if a subsequent vaccine were clinically superior, the price paid would be related to the marginal improvement the subsequent vaccine represents over the original vaccine, and the original vaccine developers would continue to receive compensation in line with the social value of their work. A bonus payment system provides a potential mechanism for doing this. One option would be to retain the exclusivity clause even if a superior vaccine were developed, but give the developer of the original vaccine incentives to buy out the technology of the second producer. The bonus payments that would go with supplying a superior vaccine would provide such an incentive. Alternatively, the newer vaccine could be purchased at a price based on its efficacy, but the developer of the newer vaccine could then be required to pay the original developer an amount equal to the price paid for the original vaccine, less an allowance related to the production cost of the new vaccine. While this approach matches private and social research incentives more closely than the blanket exception for superior products in the Orphan Drug Act, it would be difficult to administer.

VI. Vaccine Coverage and Pricing

This section first argues that the key determinant of research incentives will be the total revenue generated by a vaccine, rather than the price per person immunized. Decisions about where it is cost-effective to vaccinate should be based on the incremental cost of manufacturing an additional unit of vaccine, rather than the average price paid per person immunized under the program. Given the desired market size and number of required vaccinations, the price per person immunized can be determined by dividing the desired market size by the number of people needing immunization. The tricky question is determining the appropriate market size. The total market promised should be large enough to stimulate research, but not so large that a vaccine purchase program would not be cost-effective. The second and third subsections, titled What Size Market Is Needed to Spur Research? and Cost-Effectiveness, note that a rough rule of thumb in the industry is that a market of $250 million per year is necessary to spur significant research, and argue that a vaccine purchase program would be highly cost-effective even at a substantially larger scale. The sponsor of a vaccine purchase commitment could start with a modest program, which would not be too expensive, but retain the option to increase the value of the program if the original program proved too small to stimulate sufficient research. The fourth subsection, titled Increasing the Promised Vaccine Price over Time, argues that as long as the vaccine price is not expected to increase too quickly, this will not lead vaccine developers to withhold a vaccine from the market in the hope of getting a better price.

Coverage

The key determinant of research incentives will be the total discounted revenue generated by a vaccine. It is very expensive to conduct research, but once research is complete, it is typically fairly cheap to produce additional doses. For a fixed amount of total revenue, vaccine developers will therefore be almost as happy to produce a high volume at a low price as a low volume at a high price.

This implies that, at least as a first approximation, prices should be set per person immunized, not per dose. There is little reason to pay more per person immunized if more doses are required to provide immunity than if a single dose is required. In fact, the vaccine is more valuable if only a single dose is required to provide immunity, as this reduces delivery costs and is likely to increase patient compliance.

Moreover, the vaccine purchase program would not save money by excluding large countries from coverage, or excluding countries if vaccination is cost-effective at the marginal cost of production, but not at the average price paid for vaccine under the program. This is a false economy, because potential vaccine developers will need a fixed amount of revenue to induce them to conduct research, and if fewer doses are purchased, the price per person immunized will need to be greater to induce the same amount of research.[21]

Given the quantity of vaccines likely to be needed, the price per immunized person should be set so as to yield the desired market size. Market size should be large enough to stimulate research if scientifically warranted, but not so large that a vaccine would not be cost-effective.

What Size Market Is Needed to Spur Research?

There is no single answer to the question of how large a market is needed to spur research. The larger the market for a vaccine, the more firms will enter the field, the more research leads each firm will pursue, and the faster a vaccine will be developed. The more researchers entering the field, the smaller the chance that any particular firm will be the first to develop a vaccine. Thus the cost of development, adjusted by the risk that a particular firm or research team will not win the development race, rises with the potential size of the market. Given the enormous burden of malaria, tuberculosis, and HIV/AIDS, it is important to provide sufficient incentive for many researchers to enter the field and to induce major pharmaceutical firms to pursue several potential leads simultaneously so that vaccines can be developed quickly.

Because potential vaccine developers know that their research may fail, in order to have incentives to conduct this work, they must expect to more than cover their research expenses if they succeed. For example, if potential biotechnology investors expect that a candidate vaccine has a 1 in 10 chance of succeeding, they would require at least a tenfold return on their investment in the case of success to make the investment worthwhile.[22]

There are several ways to get a sense of the minimum market size needed to motivate investors. DiMasi et al. (1991), who examined 93 randomly selected new chemical entities from a survey of 12 pharmaceutical firms and found that, taking into account the risk of failure at each stage in the drug development process, the average cost per approved New Chemical Entity (NCE) was $114 million 1987 dollars. Capitalizing this to the date of marketing approval at a (probably overgenerous) 8% discount rate implies an average cost of $214 million 1987 dollars, or approximately $313 million 1999 dollars. While this figure is of some interest, there is wide variation in the cost of developing pharmaceuticals. DiMasi found that for most stages in the vaccine development process, the standard deviation of cost was greater than the mean cost. Vaccine trials for diseases with low incidence, such as HIV and tuberculosis, require very large samples, and are therefore expensive.[23]

The cost of developing malaria, tuberculosis, or HIV vaccines may be much higher than suggested by these estimates, since surveys of existing drugs and vaccines are disproportionately likely to focus on the low hanging fruit of entities that are cheap to develop. Unfortunately, vaccines for malaria, tuberculosis, and HIV may not be such low hanging fruit.

It is also useful to consider the revenue streams which seem sufficient to induce vaccine research in developed countries. The new Varivax vaccine against chickenpox is expected to average about $177 million in annual revenue for the first 7 years of its sales (Merck Annual Report 1998).

One approach to estimating the necessary size of a program is to ask pharmaceutical executives whether a vaccine purchase program could serve as an important incentive for research, and how big the program would need to be to do so. There are several reasons why this approach may give misleading results. First, the question is misspecified. As discussed above, firms must decide not merely whether to invest in developing a particular vaccine, but also at what level to invest. The more lucrative a market, the more leads they will pursue. Second, pharmaceutical executives may see the question as part of a price negotiation, and may therefore inflate their estimates, particularly if they expect that budgets are likely to be cut in a process of negotiation. Third, pharmaceutical firms may well request programs that increase their profits, without necessarily increasing their incentives to develop a new vaccine. In particular, pharmaceutical executives may claim that the most useful motivator for HIV vaccine research would be higher prices on existing vaccines. Pharmaceutical executives clearly have an incentive to claim this, whether or not it is the case. Fourth, pharmaceutical firms have been criticized for failing to invest in research on vaccines for diseases that kill millions of people, while investing in more commercially viable drugs (Silverstein 1999). This may make executives reluctant to admit that they are not investing in vaccines because they think such vaccines would not be profitable. It is more politically acceptable for executives to say that they are not investing because they see few scientific prospects for such a vaccine. Finally, the key decision makers are not just pharmaceutical firms, but also biotech firms and their potential investors. Scientists working on vaccines may not have even considered the possibility of starting biotech firms or seeking investors, but if a large market were expected for vaccines, they might start thinking about this. Given that they probably have not spent that much time thinking about these issues yet, their responses to questions may not be that informative.

The opinion of outsiders familiar with the industry but not part of it may be somewhat more credible. A respected pharmaceutical consulting firm estimates that a $250 million annual market is needed to motivate pharmaceutical firms (Whitehead 1999). A 10 year purchase commitment would likely be sufficient to motivate research, given that potential vaccine developers are likely to heavily discount sales after this period, and that competing vaccines are likely to emerge after 10 years in any case, and drive down prices to the point at which they could be more broadly affordable.[24] A condition of participation in the program could be agreement to license the vaccines to producers in developing countries after 10 years of purchases at an appropriate level.

If politicians are unwilling to assume liability for more than a fixed amount of potential expenditure, coverage under the program could be capped. For example, suppose that a $250 million annual market was deemed necessary to spur serious research on each vaccine, but that political leaders were unwilling to commit to more than $520 million in potential annual expenditures on new vaccines. Suppose also that the chance that malaria, HIV, and tuberculosis vaccines were all developed simultaneously was judged to be less than 10%. Instead of only covering vaccines for two diseases, an alternative approach would be to pledge $260 million in annual purchases for vaccines for any of the diseases, subject to a $520 million cap on total committed annual expenditures. In the unlikely case that vaccines for three diseases were developed simultaneously, purchases for each would average one-third of $520 million or $173 million. The expected market for a vaccine developer would be $0.9 \times \$260$ million $+ 0.1 \times \$173$ million, or $251.3 million.

Cost-Effectiveness

While the need to motivate research sets a lower bound on the size of the purchase commitment, the need for the program to be cost-effective when compared to alternative health interventions sets an upper bound on the size of a purchase commitment. This section argues that given the level of funding which is likely to be forthcoming, this is unlikely to prove a serious constraint. The World Bank has defined health interventions that cost less than $100/DALY saved as highly cost-effective (World Bank 1993). A program to purchase vaccines for malaria, tuberculosis, and HIV would be one of the most cost-effective health interventions in the world.

Glennerster and Kremer (2000) consider preliminary estimates of the cost-effectiveness of commitments to purchase vaccines at various funding levels, vaccine efficacy levels, and required numbers of doses.

We focus on a base case of an 80% effective one-dose vaccine that could be delivered with the EPI package. The average annual market is taken to be $336 million for each vaccine, with donors contributing approximately $250 million annually, and copayments providing the remainder. The DALY burden of malaria, tuberculosis, and HIV is taken from the World Health Report (WHO 1999a). We assume coverage rates of 75% for targeted new cohorts, 50% for young children and pregnant women, and 30% for other existing cohorts. Marginal delivery costs are assumed to be $1, $3, and $5 for these groups.

We find that in the first 10 years of the program, it would be cost-effective to vaccinate approximately 600 million people against malaria, 1.7 billion against tuberculosis, and 1 billion people against HIV. The net present value of expenditures per discounted DALY saved over a 10 year horizon would be $18 for malaria, $33 for tuberculosis, and $10 for AIDS, including delivery costs. However, the benefits of the program will continue beyond the 10 year life of the purchase commitment, as competing vaccines appear and prices fall. With competition the price of the vaccine is likely to fall to a level which is affordable for governments and agencies such as UNICEF. The long run net present value of expenditures per discounted DALY saved would be $9 for malaria, $21 for tuberculosis, and $5 for AIDS. Overall, the cost would be about $10/DALY. These numbers are very rough and should simply be taken as indicating orders of magnitude, but they do suggest that vaccine purchases would be highly cost-effective relative to the $100 per DALY World Bank threshold. Dividing the $336 million annual market by the required number of doses yields a vaccine price per person immunized in the first 10 years of $5.38 for malaria, $2.03 for tuberculosis, and $3.43 for HIV.

Purchase commitments would remain cost-effective under a range of alternate assumptions about vaccine efficacy, the number of vaccine doses required, and the size of the fund. In particular, even if vaccine efficacy were only 30%, immunization coverage for new cohorts was only 50% rather than 75%, the overall fund size was $500 million per disease per year, or three doses were needed, the program would remain cost-effective.

These estimates are likely to be conservative, as we have not taken into account some important but difficult to quantify effects. (1) Immunization programs are likely to reduce secondary infections, particularly for HIV and tuberculosis. (2) We have assumed that the population and prevalence of the diseases are at steady state. Given the fixed costs of research and development, population growth will tend to reduce the price per immunization and the cost-effectiveness of the program. (3) HIV prevalence is growing, which would lower the cost of the program per DALY saved. (4) It is possible that with widespread immunization, the diseases would be eradicated, at least in some regions. In this case the benefits of the program would continue, while the delivery and manufacturing costs would fall. (5) We have neglected benefits flowing to rich countries, which are important for HIV/AIDS and tuberculosis. (6) We have assumed reasonably high manufacturing and delivery costs. (7) We have not allowed for any targeting of vaccine delivery to areas of particularly high prevalence within countries, which would improve cost-effectiveness.

Increasing the Promised Vaccine Price over Time

The sponsor of a vaccine purchase commitment program could start with a relatively modest program. If additional incentives were judged necessary to spur vaccine research, the promised price could be increased until a vaccine were developed or the price reached the social value of a vaccine.[25] This procedure mimics auctions, which are typically efficient procurement mechanisms in situations in which production costs are unknown.[26]

As long as the price promised for a vaccine does not increase at a rate greater than the interest rate, firms will not have an incentive to sit on a vaccine they have developed while waiting for the price to rise. To see this, note that a firm that delays selling a vaccine postpones its returns into the future, and therefore has to discount these returns at the interest rate. In addition, delay risks the possibility that a competitor will introduce an alternative vaccine. Finally, if the vaccine developer has already taken out a patent, delay uses up the patent life.

If the price promised to vaccine developers were increased, this increase could potentially be restricted to vaccines which were based on patents that had not yet been taken out. Greater incentives may not be needed to stimulate the final stages of research on a candidate vaccine that is already promising. Moreover, restricting price increases to vaccines based on new patents reduces the chance that firms will withhold a product from the market in the hope that prices will increase. Pharmaceutical firms are not likely to risk delaying patent applications for fear that a competitor will preempt them, especially since there are potentially many competing biotech firms that could patent vaccines, whereas only a few large pharmaceutical firms actually conduct clinical trials and manufacture vaccines.[27] As discussed in the appendix, increasing the price over time may induce firms to delay starting research on a vaccine, or slow down the pace of this research, but this strategic delay will not be severe if many firms can potentially compete to develop a vaccine. Moreover, while vaccine trials could not be conducted secretly, research toward patents could be, and this would make it much more difficult for potential vaccine developers to collude to increase the price by delay.

The appendix uses techniques from the economic theory of auctions to examine the effect of increases in price on vaccine development. The main results are as follows: If there are many competing firms, a system in which the price starts low and rises over time will generate a vaccine at close to the lowest possible cost. The fewer competing researchers, the longer each waits before beginning vaccine research. The greater the initial price, the more rapidly a vaccine will be developed. This implies that if society values a vaccine highly, it should choose a high initial price, and thus be willing to incur the risk of paying more than the minimum cost necessary to spur vaccine development. In the most realistic case, increasing the growth rate of the price will speed vaccine development unless very few firms could potentially compete to develop the vaccine.

VII.   The Scope of a Purchase Commitment

Potentially, advance purchase commitments could be used to encourage research not only on vaccines, but also on other techniques for fighting disease, including drugs, diagnostic devices, and insecticides against the mosquitoes that transmit malaria.

Covering a range of technologies would avoid biasing research effort toward vaccines, rather than other technologies to fight disease. The example of the British government's prize for a method of determining longitude suggests that prize terms should be set so as to admit a variety of solutions. Most of the scientific community believed that longitude could best be determined through astronomical observations, whereas the actual solution was through development of a sufficiently accurate clock. Prespecifying an astronomical solution would have been a mistake.

On the other hand, opening up the program to any method of fighting disease would make defining eligibility and pricing decisions almost impossible. For example, developers of new HIV counseling techniques could seek to obtain payments for new techniques for promoting safe sex. Resources would be wasted in disputes over the impact of such programs. If only vaccines for malaria, tuberculosis, and HIV were eligible, the resources wasted on administration and on attempts to influence the committee would likely be fairly small relative to the cost of developing a vaccine, since only those who had actually developed a vaccine would have an entry ticket to begin trying to influence the disposition of program funds. One factor that militates toward restricting the program to vaccines and drugs is that existing institutions, such as the U.S. FDA, already have a reputation for adjudicating safety and efficacy of vaccines and drugs. A safe, environmentally appropriate insecticide might be an excellent way to fight malaria, but a whole set of procedures would need to be developed to determine eligibility standards for insecticides. This suggests that research on insecticide might be better supported through push programs.

In principle, purchase commitments are appropriate for both drugs and vaccines, but if a choice has to be made for budgetary reasons, vaccines are probably a slightly higher priority, since distortions in vaccine markets are more severe. Since drugs are much more susceptible than vaccines to the spread of resistance, individual decisions to take drugs may potentially create negative, as well as positive, externalities. Moreover, drugs are widely considered to be more profitable than vaccines, perhaps because consumers are reluctant to spend on vaccines for either behavioral or learning reasons.

Table 3.1 shows the number of deaths caused annually by various diseases for which vaccines are needed. Given a sufficient budget, it might be appropriate to commit in advance to purchase vaccines developed against any of these diseases. However, if funding is tightly limited, it may be appropriate to target the most deadly diseases. An alternative option would be to start with some easier-to-develop

Table 3.1
Deaths from Disease for which Vaccines Are Needed

Diseases	Deaths (000)[a]	%
AIDS	2285	27.47
Tuberculosis	1498	18.01
Malaria	1110	13.34
Pneumococcus[b]	1100	13.22
Rotavirus	800	9.62
Shigella	600	7.21
Enterotoxic E. coli	500	6.01
Respiratory syncytial virus[c]	160	1.92
Schistosomiasis[d]	150	1.80
Leishmaniasis	42	0.50
Trympanosomiasis	40	0.48
Chagas disease	17	0.20
Dengue	15	0.18
Leprosy	2	0.02
Total deaths	8319	100.00

[a]Estimated, World Health Report (WHO 1999a).
[b]A pneumococcus vaccine was just approved for use in the United States, but it needs to be tested in developing countries, and perhaps modified accordingly.
[c]The Jordan Report (NIAD 1998).
[d]R. Berquist, WHO, personal communication.
Source: Children's Vaccine Initiative. 1999, July. *CVI Forum* 18: 6.

vaccines and drugs as a way of building credibility. It also may be useful to first experiment with purchase commitments for a few vaccines or drugs and then consider modifying or extending the program based on the resulting experience.

VIII. Conclusion

For a vaccine purchase commitment to stimulate research investment, it must provide a credible promise that developers of good vaccines will be rewarded. Eligibility requirements could include both minimal technical standards and the market test that developing countries be willing to provide a copayment for the vaccine. To provide incentives for development of high quality vaccines, bonus payments for vaccines could be tied, directly or indirectly, to the number of lives or DALYs saved by the vaccine, and to the delivery cost. The developer of the first viable vaccine could have market exclusivity unless subsequent vaccines are clinically superior. The vaccine price promised per immunized child could initially be set at a modest level, and could then be raised if it proved insufficient to spur enough research.

This conclusion briefly discusses the politics surrounding vaccine purchase programs. It then discusses the proposed U.S. tax credit for qualifying vaccine sales and the proposed World Bank $1 billion fund for purchasing vaccines for future diseases. Finally, it discusses how a private foundation could implement a vaccine purchase commitment.

The Politics of Creating Markets for Vaccines

Those with a stake in current aid programs and in grant funded research programs may object to pull programs designed to create markets for vaccines, if they fear that resources would be drawn from important existing initiatives. Organizations involved in efforts to encourage condom use, for example, may fear that funds to develop an AIDS vaccine would be drawn from prevention efforts. Academic and government scientists working on HIV research may be concerned that a vaccine purchase program may result in cuts in other important research programs. These groups are well placed to affect the political decision-making process.

Conflict between the need for incentives to develop new vaccines and existing prevention and research efforts will be limited if a purchase commitment is financed from pledges rather than current budgets. When a vaccine became available, it might be seen as justifying increasing the total aid budget. Alternatively, once a vaccine became available, some existing prevention efforts may be less cost-effective, and budget savings will be possible. The prospect of these future cuts will be politically easier than cutting existing programs, as future aid budgets do not have as much constituency among aid workers as current aid budgets. The people currently promoting condom use or researching HIV may have retired or gone on to other jobs by the time an HIV vaccine has been developed. It is worth noting that the budgetary conflict between research on new vaccines and efforts to control disease using existing technologies is sharper if research is financed out of current budgets, as it would be in push programs, than if it is financed through future vaccine purchases, which would come out of future budgets.

At least in the U.S. Congress, pharmaceutical firms are also likely to be a key player in discussions of how to encourage vaccine research and development. Pharmaceutical firms will be interested in seeing some expenditure early in the program. This may be in part because such expenditures would enhance the credibility of the commitment, and in part because a program rewarding, say, a malaria vaccine, would not necessarily yield high expected profits, since much of the profit would be dissipated in competition to develop the vaccine. It may be easier to find champions for such programs in the pharmaceutical industry if some portion of the funds can be used to cover vaccines which are closer to development. In particular, several new pneumococcus vaccines are expected to be developed soon. Additional work will be needed to test the suitability of these vaccines for developing countries, and perhaps to modify them to reflect the strains of pneumococcus prevalent there. As currently written, the U.S. administration's proposal would cover new pneumococcus vaccines, since the disease kills more than a million people each year. Note, however, that one vaccine for pneumococcus has been licensed recently, and that under the administration's proposal, this particular vaccine would not be eligible, since it was developed before the legislation was passed.

Potential Sponsors of New Markets for Vaccines

Commitments to purchase vaccines could be undertaken by governments of industrialized countries, the World Bank, or private foundations. One institution could establish the basic infrastructure for a program and make an initial pledge and other organizations could later make pledges of their own. The initial pledge could cover particular diseases or countries, with later pledges broadening the program. Nations might not want to pledge to a vaccine purchase commitment program operated under another donor nation's control, so it might make sense to build in procedures for representation of multiple donors on decision-making bodies at the start, even if the program were initially supported by only one or two donors.

The U.S. administration's 2000 budget proposal (available at http://www.treas.gov/taxpolicy/library/grnbk00.pdf) included $1 billion in tax credits on vaccine sales over the 2002–2010 period. The program would match every dollar of qualifying vaccine sales with a dollar of tax credit, effectively doubling the incentive to develop vaccines for neglected diseases. Qualifying vaccines would have to cover infectious diseases which kill at least one million people each year, would have to be FDA approved, and would have to be certified by the

Secretary of the Treasury after advice from the U.S. Agency for International Development. To qualify for the tax credit, sales would have to be made to approved purchasing institutions, such as UNICEF. Although the President's proposal is structured as a tax credit, it would have effects similar to an expenditure program that matched private funds spent on vaccines. The administration's proposal could help catalyze other funds for vaccine purchases, since it matches such purchases dollar for dollar.

The details of which vaccine sales would qualify would be worked out by the U.S. Agency for International Development (USAID) under the program, and the analysis in this paper suggests that the details of their procedures will be quite important for the effect of the program. Biotech and pharmaceutical firms are more likely to find the commitment credible if, once the tax credit legislation is passed, USAID quickly specifies guidelines for how it will allocate credits. In particular, USAID would need to specify how it will address issues of vaccine pricing (presumably, it would not approve credit allocations for a small quantity of vaccine sold at tens of thousands of dollars per person immunized); how much of the fund could be spent on a vaccine that is currently far along in research, such as the pneumococcus vaccine; and what procedures would be used to allocate credits if multiple versions of a vaccine were available.

The World Bank president, James Wolfensohn, recently said that the institution plans to create a $1 billion loan fund to help countries purchase specified vaccines if and when they are developed (Financial Times 2000). Glennerster and Kremer (2000) discuss this proposal in more detail. The Bank has yet to take action on this. One option under consideration is a more general program to combat communicable diseases of the poor. For a general program to stimulate research, it must include an explicit commitment to help finance the purchase of new vaccines if and when they are developed. Without an explicit commitment along the lines proposed by Wolfensohn, it is unlikely that the large scale investments needed to develop vaccines will be undertaken. As discussed in the companion paper, increased coverage of existing vaccines, while desirable in its own right, will by itself be inadequate to convince potential vaccine developers that there will be a market for new vaccines when they are developed, given the long lead times for vaccines and the fickleness of donor interest.

An explicit commitment to help finance purchases of new vaccines will not interfere with other initiatives to tackle communicable diseases of the poor. This is because the commitment does not have to be financed unless and until a vaccine is developed. So, for example, the Bank could increase lending to promote the use of bednets against malaria, or increase coverage of existing vaccines, while committing that if and when new vaccines are developed, it will provide loans to countries purchasing these vaccines.

Some within the Bank have traditionally regarded earmarking future credits for a particular purpose as undesirable because it reduces the flexibility of the Bank to provide loans where they would achieve the greatest benefit. Sacrificing flexibility is a mistake when it brings no compensating advantage. However, earmarking can be justified as a response to time consistency problems. In particular, in the case of vaccines, earmarking can help resolve the time consistency problem inherent in convincing potential vaccine developers that governments will compensate them adequately once they have sunk funds into developing vaccines. The loss of flexibility associated with earmarking does not seem like a major problem, since it would be hard to imagine a situation in which purchasing vaccines for malaria, tuberculosis, and AIDS would not be cost-effective. In any case, a commitment could be structured so that it would be triggered only if a vaccine satisfied a particular cost-effectiveness threshold.

For countries to have an incentive to participate in the proposed World Bank program, loans will need to be at the concessional International Development Association (IDA) rates, and must not simply substitute for other concessional loans countries would have received. This is because commitments by one country to purchase vaccines benefit other countries by encouraging vaccine research and development. No one country, therefore, has a sufficient incentive to make a commitment on its own (the global public good problem).

Private foundations could also play a major role in creating markets for new vaccines. Foundations may find it easier than governments to credibly commit to future vaccine purchases, given their greater continuity of leadership. In particular, the Gates Foundation has $22 billion in assets, and one of its main priorities is children's health in developing countries, and vaccines in particular. U.S. law requires private foundations to spend at least 5% of their assets annually. This suggests a way that push and pull incentives for vaccine development could be combined. A U.S. foundation could spend 5% of its assets annually on grants to help expand the use of existing vaccines and provide for vaccine research. Meanwhile, the foundation could put its principal to use in encouraging vaccine research, simply by pledging that if a vaccine were actually developed, the foundation would purchase and distribute it in developing countries.

Appendix: The Effect of Increasing the Promised Price for Vaccines

This appendix analyzes the effects of increasing the price pledged for a vaccine under the simplest model of auctions, in which each firm has a private cost of developing a vaccine, and these costs are independent. Suppose that the cost of developing a vaccine for pharmaceutical firm i, denoted c_i, is independently drawn from a distribution F with upper support $\bar{p}$ and that there are N symmetrical pharmaceutical firms. Suppose the price p starts at some value $\underline{p} < \bar{p}$ and then grows, or is expected on average to grow, at a constant rate until a vaccine is invented, or until p reaches $\bar{p}$.

An equilibrium consists of a function $p_i(c_i)$ mapping each firm's cost into a price at which it will develop a vaccine. A necessary first order condition for $p_i(c_i)$ to be privately optimal is that the growth rate of surplus, $p_i - c_i$, must equal the discount rate plus the hazard rate that a rival firm will develop the vaccine. In the simplest case, in which bidders are symmetric and the cost of developing a vaccine is not correlated among bidders, p_i increases monotonically with c_i. Given monotonicity, the hazard rate that a rival will enter depends on the probability that a rival firm has a cost slightly greater than c_i conditional on no firm having a cost less than c_i. As the number of firms grows, $p_i(c_i)$ declines, asymptotically approaching c_i, and the hazard rate that a rival enters grows without bound. Thus, if there were many symmetric pharmaceutical firms, this auction mechanism would lead a vaccine to be developed at a price very close to the cost of its development. Increasing the number of bidders not only reduces the expected price, but also reduces the expected time until a vaccine is developed given F and the growth rate of p.

At least over some range, increasing the growth rate of p, taking $\underline{p}$ as fixed, will speed the time until a vaccine is developed. This is despite the fact that the first order condition implies that the faster the growth rate of p, or equivalently the lower the discount rate, the greater $p_i(c_i)$. To see why increasing the growth rate of p speeds the auction, note that if the growth rate of p is infinite, then the auction concludes immediately because the price immediately attains its upper limit of $\bar{p}$. As the growth rate of p approaches zero, the expected time for the auction to conclude grows without bound. Moreover, reducing the growth rate of p must asymptotically increase the time until a vaccine is developed, since as $\dot{p}/p$ approaches zero, $p_i(c_i)$ approaches its lower bound of c_i, and hence as the growth rate slows, the reduction in p_i is bounded, whereas the time it takes for the auction to reach any particular price increases without bound as the auction slows.

It seems likely that the expected time until a vaccine is produced typically declines with the growth rate of p, given $\underline{p}$, but if there are few firms, it is possible to construct examples in which the expected time until a vaccine is produced increases with the growth rate of p. If there are many firms, then $p_i(c_i)$ will be very close to c_i, and hence reducing the growth rate of p will have little effect on $p_i(c_i)$, but will still lengthen the time required to reach any price. Hence, with many firms, a rapidly growing price, given $\underline{p}$, is likely to lead to a much faster vaccine discovery. On the other hand, if there are only a small number of firms, then $p_i(c_i)$ may be significantly greater than c_i, and reducing $p_i(c_i)$ may significantly shorten the auction. Consider the extreme case with only one firm. If p grows rapidly enough, the bidder will prefer to wait until the end of the auction, when the price reaches $\bar{p}$, before developing a vaccine. On the other hand, if the growth rate of the price is less than the interest rate, then once p/c_i is great enough, the vaccine will be developed. Thus, at least for some realizations of c_i, increases in the growth rate of p can lengthen the time until a vaccine is developed. If the distribution of the cost of development is such that most of the mass is at a low level, but there is a thin tail reaching up to $\bar{p}$, then increases in the growth rate of p can lengthen the expected time until a vaccine is developed.

Holding constant $\bar{p}$ and the growth rate of the price, the higher $\underline{p}$, the shorter the time until a vaccine is developed. This suggests that the more a vaccine is valued, the greater $\underline{p}$ should be. In the extreme, if the social value of the vaccine is far greater than the upper support of c, then it would make sense to either have the price rise very quickly, or to choose $\underline{p}$ close to $\bar{p}$. Some may feel that the social value of vaccines is so great that it is better to spend more money than to risk delay, but this does not seem to be the revealed preference of rich country governments.

As long as the price does not grow that much faster than the interest rate, pharmaceutical firms will not actually sit on a vaccine they had already developed, waiting for the price to rise. Given discounting, it would be better for the firm to wait to begin research, rather than to first incur the cost of developing a vaccine, and then sit on the vaccine. Even if the firm got lucky and developed a vaccine faster than it expected, it would not sit on it if the growth rate of the program were equal to or less than the discount rate. Once a vaccine is developed, the opportunity cost of losing out to another bidder is not $p - c_i$, but rather p. The firm would only wait to develop the vaccine if the growth rate of p exceeded the discount rate plus the hazard rate that another firm would develop a vaccine.[28]

The optimal initial price depends on the expected cost of developing the vaccine, and therefore would generically differ between diseases. To see this, consider a hypothetical example in which each pharmaceutical firm faces its own cost of developing a vaccine, but it is common knowledge that the cost of developing a malaria vaccine is such that research would be profitable at between $5 and $6 per person immunized, while the cost of developing an HIV vaccine is such that research would be profitable at between $15 and $16 per person immunized. Starting the auction at more than $6 per person immunized would provide unnecessary rents to developers of a malaria vaccine. Starting the auction at less than $15 per person immunized would unnecessarily delay the development of an HIV vaccine.

The analysis above treats the cost of developing a vaccine as independently distributed across bidders, but in practice, there are almost certainly common components to this cost, and to the benefits of selling a vaccine to the program. This will create some tendency toward a winner's curse. Firms might try to publicize any leads in research in order to deter rivals. This is a general feature of patent races, and is not specific to this mechanism. Since developing a vaccine involves many stages of research, and promising vaccines can fail at any stage from laboratory tests to animal trials to Phase 4 human trials, potential rivals are unlikely to believe that the leader has a lock on becoming the first to develop a vaccine.[29]

Notes

I am grateful to Daron Acemoglu, Philippe Aghion, Martha Ainsworth, Susan Athey, Abhijit Banerjee, Amie Batson, David Cutler, Sara Ellison, Sarah England, John Gallup, Chandresh Harjivan, Eugene Kandel, Jenny Lanjouw, Sendhil Mullainathan, Ariel Pakes, Ok Pannenborg, Sydney Rosen, Andrew Segal, Scott Stern, and especially Amir Attaran, Rachel Glennerster, and Jeffrey Sachs for very extensive comments. Amar Hamoudi, Jane Kim, and Margaret Ronald provided excellent research assistance. This paper is part of a Harvard Center for International Development project on vaccines. I thank the National Science Foundation and the MacArthur Foundation for financial support. The views expressed in this paper are my own, and not theirs. Department of Economics, Littauer 207, Harvard University, Cambridge, MA 02138; mkremer@fas.harvard.edu.

1. The credibility of the vaccine purchase commitment can be increased by framing it as a unilateral contract (i.e., one not requiring a promise by others to become valid) and explicitly including a promise not to revoke. Some additional legal issues might arise if a purchase commitment were made by a national government or an international institution, and legal research would be needed to address these issues.

2. Sobel (1995) argues that the longitude prize committee was biased toward an astronomical solution and insisted on improvements and multiple trials, creating repeated delays, until the king intervened on behalf of the chronometer's inventor. Note, however, that this account is disputed. The economic historian Paul David argues that the conditions imposed by the committee were reasonable (personal communication 2000). In any case, this points to the importance of program rules and adjudication procedures in influencing credibility of purchase commitments.

3. For example, consider a simple case in which potential vaccine developers seek to maximize expected profits and accurately interpret the degree of commitment entered into by potential donors. Suppose that in the absence of a particular piece of contractual language in the vaccine purchase commitment, there is a 90 percent chance the sponsor purchases the vaccine at the promised price and a 10 percent chance that they renege and renegotiate to a price of half the level originally promised. In this case, in the absence of a contractual arrangement, firms which seek to maximize expected profits will act as if the value of the program is not the promised annual revenue from the program, but rather 95 percent of the promised annual revenue. Note that while the expected incentive is only 95 percent of the promised level, so is the expected cost to the sponsor. To the extent that both vaccine developers and the sponsor are risk averse, they would both prefer a perfectly credible commitment of $950 million to a 90 percent chance of $1 billion and a 10 percent chance of a $500 million payment. In this sense, imperfect credibility reduces the efficiency of purchase commitments.

4. Some have speculated about the possibility of an altruistic malaria vaccine, which would block further transmission of the disease, without protecting the person who takes the vaccine. It is unclear how many people would be willing to take such a vaccine. Moreover, given the high intensity of malaria transmission in many parts of Africa, the epidemiological impact of an altruistic vaccine might be quite small unless the vaccination rate was very high. Committing in advance to purchase such a vaccine would be difficult.

5. Glennerster and Kremer (2000) examine the cost-effectiveness of vaccines with different degrees of efficacy, requiring different numbers of doses, and providing different lengths of protection. In future work, we plan to extend this analysis to examine how eligibility standards could be established so that vaccines would be eligible if they meet a cost-effectiveness threshold.

6. Setting efficacy requirements for eligibility for an HIV vaccine is particularly difficult. Because of the key importance of a core group of high-risk people in influencing the spread of HIV, even a vaccine of low efficacy may prove useful in disrupting the chain of transmission if it is targeted to this group. On the other hand, at least theoretically, an imperfectly effective HIV vaccine could increase the spread of HIV, since people might adopt riskier behaviors if they felt they had reduced the chance of infection by taking an imperfect HIV vaccine. This outcome seems unlikely, however, since in steady state, an imperfectly effective vaccine could also potentially make the highest activity people

more hopeful about their chances of being uninfected, and therefore less likely to adopt risky behavior. Delivery of an HIV vaccine may have to use very different channels than delivery of existing childhood vaccines, particularly if it is targeted to such a core group. Little is known about the costs of reaching such groups.

7. Note that the problem of inducing firms to conduct research and development on vaccines for which they expect the government to be the major purchaser is in some ways similar to the problem of inducing firms to conduct research and development on weapons for which they expect governments will be the major purchaser. In each case, the government must convince the firms contemplating undertaking research that it will not take advantage of them by insisting on low prices once they have already sunk their investments in research. Procurement rules for the U.S. Department of Defense do not instruct procurement officers to purchase orders at the lowest possible price, but instead to purchase at a price that covers suppliers' costs. The formulas used for calculating costs typically allow firms to cover more than manufacturing costs, which in turn provides an incentive for firms to invest in research and development to produce attractive products that allow them to win procurement contracts. Rogerson (1994) suggests that this serves as a reputational mechanism for encouraging research by defense contractors. The Defense Department has an advantage in that it is a long-standing institution, with a well developed reputation about how it treats contractors, and contractors can count on the desire of the Defense Department to maintain a reputation for the future, because the continued existence of the Defense Department seems assured. Unfortunately, the long-term future of a vaccine purchase program is less certain.

8. Unfortunately, there is a history of antagonism between the pharmaceutical industry and existing international vaccine purchasers such as the Pan American Health Organization (PAHO) and the United Nations' Children's Fund (UNICEF), which have a culture of trying to purchase vaccines at the minimum possible price. These institutions, therefore, might have difficulty administering a program designed to increase private sector incentives for vaccine development.

9. On the other hand, if the program maintained a single fund which could be used to purchase vaccines for any of several different diseases, then potential vaccine developers might fear that once they had invested money in developing a vaccine, the vaccine purchase program would try to pay a very low price for the vaccine, hoping to save its resources to purchase vaccines for other diseases. This problem could be addressed by maintaining separate funds (or making separate financial commitments) for different diseases.

10. Setting low prices is the most likely way that the program could take advantage of vaccine developers. Program adjudicators concerned with public health will have limited incentives to insist on further trials, for example, because they will presumably want to get an effective vaccine into the field.

11. This is illustrated vividly by the apparently meager prospects of the Wyeth-Ayerst rotavirus vaccine in developing countries after it was withdrawn from the U.S. market following evidence that it causes intussusception in rare cases. The benefits of the vaccine are likely to outweigh by far its risks in developing countries, where rotavirus kills three-quarters of a million children each year. Nonetheless, it appears unlikely that the vaccine will ever be widely used.

12. Willingness to pay is also likely to be higher for countries with a greater burden of disease, but requiring a larger co-payment from countries with a greater disease burden seems inequitable and is likely to be politically infeasible.

13. It might therefore, for example, be appropriate to specify that the program could require proof of efficacy over some extended period for sporozoite malaria vaccines.

14. If interest were paid on accounts, countries would be under less time pressure to reach agreement with vaccine developers, and therefore might have such a strong bargaining position that they could prevent vaccine developers from recovering their research costs. Note that vaccine developers are automatically under time pressure to reach a deal with purchasers, because their patent is time limited. Moreover, if interest is not paid on individual country accounts, then any interest accumulated on the program could be used to fund grants for basic vaccine research, or allocated to countries where disease prevalence had increased since the program was established.

15. Payments by third parties are also difficult to regulate. Suppose a Swiss firm invents a malaria vaccine which is not effective against the strains of malaria prevalent in some country, and therefore is not appropriate for that country. The government of Switzerland or a foundation supported by the firm could provide aid for purchasers to use towards their copayments. With a 20% copayment, this would allow the government of Switzerland or the foundation to spend 1 dollar to raise 5 dollars for the company.

16. It is worth noting that currently, the medical profession and society as a whole seem to weight DALYs caused by side effects much more heavily than DALYs saved.

17. Information about the number of lives or DALYs saved might become available only gradually, and therefore, if this approach were adopted, it might theoretically be best to condition payments on long run outcomes. For example, it might initially be unclear whether a vaccine provides protection only temporarily, or indefinitely. The extent to which a vaccine prevents secondary infections might also be difficult to predict in advance. Initial bonus payments to vaccine developers could be based on conservative estimates of lives or DALYs saved and additional payments could be made later, depending on the realization of lives or DALYs saved. Of course, if payments were delayed, accumulated interest would have to be paid as well. Basing bonus payments to vaccine developers on realized DALYs or lives saved, rather than on the results of the clinical trials required for regulatory approval, creates better incentives to develop vaccines that will work in the real world, rather than only in clinical trials, where it is easier to make sure that delivery protocols are followed exactly. Moreover, if bonus payments could be claimed after a vaccine had already been used, it would be much more difficult for a price setting committee within the vaccine purchase program to refuse to pay a remunerative price. Before a vaccine is used in the field, the committee could argue that it deserves only a small bonus, citing potential problems with the vaccine. However, if the vaccine is used, and it reduces the burden of malaria by 90%, it will be very hard for the committee to argue that it is ineffective. (Exceptions to this are new diseases, such as HIV, for which predictions of prevalence in the absence of a vaccine are likely to be particularly inaccurate.)

18. Basing incentives on mortality rather than DALYs might be attractive, since mortality is easier for the public to understand and perhaps less subjective and open to manipulation. On the other hand, it may be best to more closely tie incentives to objectives by rewarding DALYs saved. It is desirable to give researchers incentives to reduce morbidity as well as mortality, and to guard against side effects that cause morbidity.

19. For example, in Africa HIV prevalence can be taken as a good indicator of future HIV deaths and disability, but prevalence of malaria may be a poor indicator of the total burden of malaria, since a vaccine might greatly reduce malaria mortality without preventing infection.

20. If the vaccine purchase program were an international organization, it is not clear what court would have authority to rule on intellectual property rights questions. One option would be to spend funds from each donor in accordance with the intellectual property rights laws of that country. For example, U.S. funds would not be used to purchase vaccines that violate U.S. patents.

21. Excluding countries that would have bought vaccine in the absence of a program at prices greater than or equal to the price paid by the program would, however, increase incentives to develop vaccines. A sliding scale of copayments could be used to gradually phase out the program.

22. As discussed in the companion paper, advocates for grant-funded research programs may have incentives to be over-optimistic about the prospects for easily developing vaccines. The Institute of Medicine estimated in 1986 that a malaria vaccine could be developed for $35 million. This estimate is far too low. From the limited description of their methodology, it seems that their cost estimate assumes success in every stage of the vaccine development process, while in fact, it is likely that many different candidate vaccines will have to be tried before a usable vaccine is developed. A further indication that the Institute of Medicine's estimates were over-optimistic lies in their 1986 prediction that a malaria vaccine could be licensed within 5 to 10 years.

23. Regulators may require large samples even for vaccines for diseases with higher incidence, because they believe it is especially important to detect potential side effects of vaccines, since they are administered to healthy people.

24. The life of a patent is 20 years. However, a vaccine would only reach the market several years after the date of application for a patent. The effective life of a patent is the number of years remaining on the patent from the time that it is first brought to market. Shulman, DiMasi, and Kaitin (1999) report that the average effective patent life for new drugs and biologicals is 11.2 years under the Waxman-Hatch Act, which granted extra protection to inventors to partially make up for loss of patent life during regulatory review. Without the Act, patent life would be 8.2 years. The Act covers the U.S. only, and there is no reason to believe that developing countries will offer similar patent protection. As noted above, a requirement to license vaccines after 10 years could potentially be built into the program.

25. Since the quantity purchased would stay constant, total revenue would rise in proportion to price.

26. Another option would be to preannounce that if no vaccine had been developed by a certain date, the price would start growing automatically. However, it is probably better to let future decision makers choose whether or not to increase the price, since in some scenarios it would be optimal not to increase the price. For example, there would be no need to increase the price if general technological advances in biology reduced the expected cost of developing a vaccine sufficiently that many firms decided to pursue vaccines.

27. One potential problem with this approach is that vaccine developers might incorporate unnecessary late-patented components in the vaccine to qualify for a higher price. However, a committee could rule on what were the key patents used in a given vaccine, so simply adding an extra useless patent would not lead to a higher vaccine price.

28. I am considering the case in which there is only one potential patented vaccine, so the winner reaps the entire reward.

29. For example, rotavirus vaccine was recently withdrawn from the U.S. market, at least temporarily, following reports of side effects.

References

Ainsworth, Martha, Amie Batson, and Sandra Rosenhouse. 1999. "Accelerating an AIDS Vaccine for Developing Countries: Issues and Options for the World Bank." World Bank, Washington, DC.

Batson, Amie. 1998. "Win-Win Interactions Between the Public and Private Sectors." *Nature Medicine* 4(Supp.):487–91.

Bishai, D., M. Lin, et al. 1999, June 2. "The Global Demand for AIDS Vaccines." Presented at the 2nd International Health Economics Association Meeting, Rotterdam.

Bos, Edwuard, My T. Vu, Ernest Massiah, and Rodolfo A. Bulatao. 1994. *World Population Projections 1994–95: Estimates and Projections with Related Demographic Statistics.* Baltimore, MD: The Johns Hopkins University Press.

Chima, Reginald, and Anne Mills. 1998, June. "Estimating the Economic Impact of Malaria in Sub-Saharan Africa: A Review of the Empirical Evidence." Working Paper, London School of Hygiene and Tropical Medicine, London.

CNNfn. 1998, May 1. "Merck Slashes Zocor Price."

CVI Forum. 1999, July. Children's Vaccine Initiative 18:6.

Department for International Development (DFID). 2000, October 19. "Harness Globalisation to Provide New Drugs and Vaccines for the Poor, Says Short." Press Release, London.

Desowitz, Robert S. 1991. *The Malaria Capers: Tales of Parasites and People.* New York: W. W. Norton.

Desowitz, Robert S. 1997. *Who Gave Pinta to the Santa Maria? Torrid Diseases in a Temperate World.* New York: W.W. Norton.

DiMasi, Joseph, et al. 1991, July. "Cost of Innovation in the Pharmaceutical Industry." *Journal of Health Economics* 10(2)107–42.

Dupuy, J. M., and L. Freidel. 1990. "Viewpoint: Lag between Discovery and Production of New Vaccines for the Developing World." *Lancet* 336:733–4.

Financial Times. 2000, February 2. "Discovering Medicines for the Poor."

Galambos, Louis. 1995. *Networks of Innovation: Vaccine Development at Merck, Sharp & Dohme, and Mulford, 1895–1995.* New York: Cambridge University Press.

Gallup, John, and Jeffrey Sachs. 2000, October. "The Economic Burden of Malaria." Working Paper, Harvard Institute for International Development, Cambridge, MA. Downloadable from http://www.hiid.harvard.edu.

GAO (General Accounting Office (U.S.)). 1999. "Global Health: Factors Contributing to Low Vaccination Rates in Developing Countries." Washington, DC.

Glennerster, Rachel, and Michael Kremer. 2000. "Preliminary Cost-Effectiveness Estimates for a Vaccine Purchase Program." Working Paper, Harvard University, Cambridge, MA.

Grosser, Morton. 1991. *Gossamer Odyssey: The Triumph of Human-Powered Flight*. New York: Dover Publications.

Guell, Robert C., and Marvin Fischbaum. 1995. "Toward Allocative Efficiency in the Prescription Drug Industry," *The Milbank Quarterly* 73:213–29.

Hall, Andrew J., et al. 1993. "Cost-Effectiveness of Hepatitis B Vaccine in The Gambia." *Transactions of the Royal Society of Tropical Medicine and Hygiene* 87:333–6.

Hoffman, Stephen L., ed. 1996. *Malaria Vaccine Development: A Multi-immune Response Approach*. Washington, DC: American Society for Microbiology.

Institute of Medicine (U.S.). 1991. Committee for the Study on Malaria Prevention and Control: Status Review and Alternative Strategies. *Malaria: Obstacles and Opportunities: A Report of the Committee for the Study on Malaria Prevention and Control: Status Review and Alternative Strategies, Division of International Health, Institute of Medicine*. Washington, DC: National Academy Press.

Institute of Medicine (U.S.). 1986. Committee on Issues and Priorities for New Vaccine Development. *New Vaccine Development, Establishing Priorities, vol. 2: Diseases of Importance in Developing Countries*. Washington, DC: National Academy Press.

Johnston, Mark, and Richard Zeckhauser. 1991, July. "The Australian Pharmaceutical Subsidy Gambit: Transmitting Deadweight Loss and Oligopoly Rents to Consumer Surplus." Working Paper no. 3783, National Bureau of Economic Research, Cambridge, MA.

The Jordan Report. 1998. Division of Microbiology and Infectious Diseases, National Institute of Allergy and Infectious Diseases, National Institutes of Health, Bethesda, MD.

Kim-Farley, R., and the Expanded Programme on Immunization Team. 1992. "Global Immunization." *Annual Review of Public Health* 13: 223–37.

Kremer, Michael. 1998, November. "Patent Buyouts: A Mechanism for Encouraging Innovation." *Quarterly Journal of Economics* 113(4):1137–67.

Kremer, Michael, and Jeffrey Sachs. 1999, May 5. "A Cure for Indifference." *The Financial Times*.

Lanjouw, Jean O., and Iain Cockburn. 1999. "New Pills for Poor People?: Empirical Evidence After GATT." New Haven, CT: Yale University.

Lichtmann, Douglas G. 1997, Fall. "Pricing Prozac: Why the Government Should Subsidize the Purchase of Patented Pharmaceuticals." *Harvard Journal of Law and Technology* 11(1):123–39.

Mercer Management Consulting. 1998. "HIV Vaccine Industry Study October–December 1998." World Bank Task Force on Accelerating the Development of an HIV/AIDS Vaccine for Developing Countries. World Bank, Washington, D.C.

Merck and Co. Annual Report, 1998. Available at http://www.merck.com/overview/98ar.

Milstien, Julie B., and Amie Batson. 1994. "Accelerating Availability of New Vaccines: Role of the International Community." Global Programme for Vaccines and Immunization. Available at http://www.who.int/gpv-supqual/accelavail.htm.

Mitchell, Violaine S., Nalini M. Philipose, and Jay P. Sanford. 1993. *The Children's Vaccine Initiative: Achieving the Vision*. Washington, DC: National Academy Press.

Morantz, Alison, and Robert Sloane. 2000, October 2. "Vaccine Pre-Payment Plan: Overview of Legal Design Issues." Memo, Harvard University, Cambridge, MA.

Muraskin, William A. 1995. *The War Against Hepatitis B: A History of the International Task Force on Hepatitis B Immunization.* Philadelphia: University of Pennsylvania Press.

Murray, Christopher J. L., and Alan D. Lopez. 1996. *The Global Burden of Disease: A Comprehensive Assessment of Mortality and Disability from Diseases, Injuries, and Risk Factors in 1990 and Projected to 2020. Global Burden of Disease and Injury Series; vol. 1.* Cambridge, MA: Published by the Harvard School of Public Health on behalf of the World Health Organization and the World Bank; Distributed by Harvard University Press.

Murray, Christopher J. L., and Alan D. Lopez. 1996. *Global Health Statistics: A Compendium of Incidence, Prevalence, and Mortality Estimates for Over 200 Conditions. Global Burden of Disease and Injury Series; v. 2.* Cambridge, MA: Published by the Harvard School of Public Health on behalf of the World Health Organization and the World Bank; Distributed by Harvard University Press.

Nadiri, M. Ishaq. 1993. "Innovations and Technological Spillovers." Working Paper no. W4423, National Bureau of Economic Research, Cambridge, MA.

National Academy of Sciences. 1996. "Vaccines Against Malaria: Hope in a Gathering Storm." National Academy of Sciences Report. Downloadable from http://www.nap.edu.

National Institutes of Health. 1999. Institutes and Offices. Office of the Director. Office of Financial Management. Funding. http://www4.od.nih.gov/ofm/diseases/index.stm.

Nichter, Mark. 1982. "Vaccinations in the Third World: A Consideration of Community Demand." *Social Science and Medicine* 41(5):617–32.

PATH (Program for Appropriate Technology in Health). At http://www.path.org.

PhRMA. 1999. PhRMA Industry Profile 1999. Available at http://www.phrma.org/publications/industry/profile99/index.html.

Pilling, David. 2000, February 2. "Discovering Medicines for the Poor." *Financial Times* 7.

Robbins, Anthony, and Phyllis Freeman. 1988, November. "Obstacles to Developing Vaccines for the Third World." *Scientific American:* 126–33.

Rogerson, William P. 1994, Fall. "Economic Incentives and the Defense Procurement Process." *Journal of Economic Perspectives* 8(4):65–90.

Rosenhouse, S. 1999. "Preliminary Ideas on Mechanisms to Accelerate the Development of an HIV/AIDS Vaccine for Developing Countries." Technical Paper, The World Bank, Washington, DC.

Russell, Philip K. 1997, September. "Economic Obstacles to the Optimal Utilization of an AIDS Vaccine." *Journal of the International Association of Physicians in AIDS Care* 3(9): 31–33.

Russell, Philip K., et al. 1996. *Vaccines Against Malaria: Hope in a Gathering Storm.* Washington, DC: National Academy Press.

Russell, Philip K. 1998. "Mobilizing Political Will for the Development of a Safe, Effective and Affordable HIV Vaccine." NCIH Conference on Research in AIDS.

Sachs, Jeffrey. 1999. "Sachs on Development: Helping the World's Poorest." *The Economist* 352(8132):17–20.

Salkever, David S., and Richard G. Frank. 1995. "Economic Issues in Vaccine Purchase Arrangements." Working Paper no. 5248, National Bureau of Economic Research, Cambridge, MA.

Scotchmer, Suzanne. 1999, Summer. "On the Optimality of the Patent Renewal System." *Rand Journal of Economics* 30(2):181–96.

Shavell, Steven, and Tanguy van Ypserle. 1998. "Rewards versus Intellectual Property Rights." Cambridge, MA: Harvard Law School. Mimeo.

Shepard, D. S., et al. 1991. "The Economic Cost of Malaria in Africa." *Tropical Medicine and Parasitology* 42:199–203.

Shulman, Sheila R., Joseph A. DiMasi, and Kenneth I. Kaitin. 1999. "Patent Term Restoration: The Impact of the Waxman-Hatch Act on New Drugs and Biologics Approved 1984–1995." *The Journal of Biolaw and Business* 2(4).

Shulman, Sheila R., and Michael Manocchia. 1997, September. "The U.S. Orphan Drug Programme: 1983–1995." *Pharmacoeconomics* 12(3): 312–26.

Silverstein, Ken. 1999, July 19. "Millions for Viagra, Pennies for Diseases of the Poor." *The Nation* 269(3):13–9.

Sobel, Dava. 1995. *Longitude.* New York: Walker and Company.

Sullivan, Michael P. 1988. "Private Contests and Lotteries: Entrants' Rights and Remedies." *American Law Reports ALR 4th* 64.

Targett, G. A. T. ed. 1991. *Malaria: Waiting for the Vaccine. London School of Hygiene and Tropical Medicine First Annual Public Health Forum.* New York: John Wiley and Sons.

Taylor, Curtis R. 1995, September. "Digging for Golden Carrots: An Analysis of Research Tournaments." *The American Economic Review* 85:872–90.

UNAIDS. 1998, December. *AIDS Epidemic Update.* Geneva.

UNAIDS HIV/AIDS Global report. 2000. www.unaids.org/hivaidsinfo/statistics/june98/global_report/index.html.

Vaccaro, Don F. 1972. "Advertisement Addressed to Public Relating to Sale or Purchase of Goods at Specified Price as an Offer the Acceptance of which Will Consummate a Contract." *American Law Reports, ALR 3rd* 43.

Wellcome Trust. 1996. *An Audit of International Activity in Malaria Research.* Downloadable from www.wellcome.ac.uk/en/1/biosfginttrpiam.html.

Whitehead, Piers. 1999. "Public Sector Vaccine Procurement Approaches: A Discussion Paper Prepared for the Global Alliance for Vaccines and Immunisation." Mimeo, Mercer Management Consulting, London.

WHO (World Health Organization) 1999a. *World Health Report 1999.* Geneva.

WHO (World Health Organization). 1999b, June 17. "Infectious Diseases: WHO Calls for Action on Microbes." Geneva.

WHO (World Health Organization) 1999c. "Meningococcal and Pneumococcal Information Page." At http://www.who.int/gpv-dvacc/research/mening.html.

World Health Organization. 1998. Report on the Global Tuberculosis Epidemic www.who.int/gtb/publications/tbrep_98/PDF/tbrep98.pdf.

WHO (World Health Organization). 1997a. *Weekly Epidemiological Report* 72: 36–8.

WHO (World Health Organization). 1997b. *Anti-Tuberculosis Drug Resistance in the World.* Geneva.

WHO (World Health Organization). 1996. *Investing in Health Research and Development: Report of the Ad Hoc Committee on Health Research Relating to Future Intervention Options.* Geneva.

WHO (World Health Organization). 1996. *World Health Organization Fact Sheet N94 (revised).* Malaria. Geneva.

WHO (World Health Organization) and UNICEF. *State of the World's Vaccines and Immunization.* WHO/GPV/96.04. Downloadable from http://www.who.int/gpv-documents/docspf/www9532.pdf.

World Bank. 1993. *World Development Report: Investing in Health.* New York: Oxford University Press for the World Bank.

World Bank. 1993. *Disease Control Priorities in Developing Countries.* Oxford Medical Publications. New York: Oxford University Press for the World Bank.

World Bank. 1993. *World Development Report 1993: Investing in Health.* Washington, DC: Oxford University Press. 1993.

World Bank. 1998. *World Development Indicators.* Geneva.

World Bank. 1999. "Preliminary Ideas on Mechanisms to Accelerate the Development of an HIV/AIDS Vaccine for Developing Countries." Geneva.

World Bank AIDS Vaccine Task Force. 2000, February 28. "Accelerating an AIDS Vaccine for Developing Countries: Recommendations for the World Bank." Geneva.

Wright, Brian D. 1983, September. "The Economics of Invention Incentives: Patents, Prizes, and Research Contracts." *American Economic Review* 73:691–707.

4

Navigating the Patent Thicket: Cross Licenses, Patent Pools, and Standard Setting

Carl Shapiro, *University of California at Berkeley*

Executive Summary

In several key industries, including semiconductors, biotechnology, computer software, and the Internet, our patent system is creating a *patent thicket:* an overlapping set of patent rights requiring that those seeking to commercialize new technology obtain licenses from multiple patentees. The patent thicket is especially thorny when combined with the risk of holdup, namely the danger that new products will inadvertently infringe on patents issued after these products were designed. The need to navigate the patent thicket and holdup is especially pronounced in industries such as telecommunications and computing in which formal standard setting is a core part of bringing new technologies to market. Cross licenses and patent pools are two natural and effective methods used by market participants to cut through the patent thicket, but each involves some transaction costs. Antitrust law and enforcement, with its historical hostility to cooperation among horizontal rivals, can easily add to these transaction costs. Yet a few relatively simple principles, such as the desirability package licensing for *complementary* patents but not for *substitute* patents, can go a long way toward insuring that antitrust will help solve the problems caused by the patent thicket and by holdup rather than exacerbating them.

I. The Patent Thicket

Is our patent system slowing down the commercialization of new technologies?

The essence of science is cumulative investigation combined with hypothesis testing. The notion of cumulative innovation, each discovery building on many previous findings, is central to the scientific method. Indeed, no respectable scientist would fail to recognize and acknowledge the crucial role played by his or her predecessors in establishing a foundation from which progress could be made. As Sir

Isaac Newton put it, each scientist "stands on the shoulders of giants" to reach new heights.

Today, most basic and applied researchers are effectively standing on top of a huge pyramid, not just on one set of shoulders. Of course, a pyramid can rise to far greater heights than could any one person, especially if the foundation is strong and broad. But what happens if, in order to scale the pyramid and place a new block on the top, a researcher must gain the permission of each person who previously placed a block in the pyramid, perhaps paying a royalty or tax to gain such permission? Would this system of intellectual property rights slow down the construction of the pyramid or limit its height?

Clearly, pyramid building, namely research and development (R&D), is taking place at an impressive pace today, so there is no great cause for alarm, especially in the area of basic research where the "royalty" is often (but not always) nothing more than a citation. As we move from pure R to applied R and ultimately to D, however, one can fairly ask whether our legal and commercial institutions are in fact properly designed to promote rather than discourage the creation of products and services that draw on many strands of innovation and thus potentially require licenses from multiple patent holders. To complete the analogy, *blocking patents* play the role of the pyramid's building blocks.

Mixing metaphors, thoughtful observers are increasingly expressing concerns that our patent (and copyright) system is in fact creating a patent thicket, a dense web of overlapping intellectual property rights that a company must hack its way through in order to actually commercialize new technology. With cumulative innovation and multiple blocking patents, stronger patent rights can have the perverse effect of stifling, not encouraging, innovation.[1]

In fact, even while a consensus has emerged that innovation is the main driver of economic growth, we are witnessing somewhat of a backlash against the patent system as it is currently operating. Especially unpopular are patents on business methods, such as Priceline.com's patent on "buyer-driven conditional purchase offers" (asserted against Microsoft) or Amazon's patent on a one click online shopping system (asserted against Barnes & Noble). The Patent and Trademark Office (PTO) does indeed seem to have allowed a number of patents on ideas that would not appear offhand to meet the usual standards for novelty and nonobviousness, such as the patent held by Sightsound.com which reputedly covers "the sale of audio or video recordings in download fashion over the Internet." Emboldened by a key appeals court decision in 1998 supporting a patent for a business method enabled by computer software, patent applications for computer-related business methods have jumped from about 1,000 in 1997 to over 2,500 in 1999. In an attempt to call a truce in what could otherwise prove to be a mutually destructive patent battle, Jeff Bezos, the Chairman of Amazon.com, recently suggested that patents on software and Internet business methods be limited to 3 or 5 years, rather than the usual 20 years from the date of application.[2]

But concerns about a patent thicket, and excessively loose standards at the PTO, are hardly confined to e-commerce and business method patents. For example, in the semiconductor industry, companies such as IBM, Intel, or Motorola find it all too easy to unintentionally infringe on a patent in designing a microprocessor, potentially exposing themselves to billions of dollars of liability and/or an injunction forcing them to cease production of key products.[3] So-called submarine patents, that take years if not decades to work their way through the Patent and Trademark Office, are another great source of anxiety, especially for large manufacturing firms. Plus, more and more companies are following the lead of Texas Instruments and engaging in patent mining, trying to get the most out of their patents by asserting them more aggressively than ever against possible infringing firms, even those who are not rivals. And considerable research shows that companies are increasingly inclined to seek patents, causing an increase in the propensity to patent, as well as an increase in the practice of defensive patenting.[4]

In short, our patent system, while surely a spur to innovation overall, is in danger of imposing an unnecessary drag on innovation by enabling multiple rights owners to "tax" new products, processes, and even business methods. The vast number of patents currently being issued creates a very real danger that a single product or service will infringe on many patents. Worse yet, many patents cover products or processes already being widely used when the patent is issued, making it harder for the companies actually building businesses and manufacturing products to invent around these patents. Add in the fact that a patent holder can seek injunctive relief, that is, can threaten to shut down the operations of the infringing company, and the possibility for holdup becomes all too real.

This paper takes as given the flood of patents currently being issued by the PTO, and assumes that these patents are indeed creating a patent thicket in the sense that many new products would likely infringe on multiple patents. Remaining agnostic (but suspicious) about whether the PTO is too lax in granting patents (especially software patents), or whether the courts are too generous in upholding patents that are granted, I look at the business arrangements that are being used to cut through the patent thicket.

More specifically, I consider the evolving and growing role of *cross licenses* and *patent pools* to solve the complements problem that arises when multiple patent holders can potentially block a given product. I discuss specifically the *standard setting process*, that increasingly involves complex negotiations over patent rights and licensing terms. I also consider other ways in which companies resolve disputes over intellectual property, including acquisitions.

For each business practice, in addition to describing the economics underlying that practice and examples of its use, I consider whether antitrust limits are contributing to the problems caused by the patent system. Unfortunately, antitrust enforcement and antitrust law have a deep rooted suspicion of cooperative activities involving direct competitors. But such cooperation, in one form or another, may be precisely what is required to navigate the patent thicket. As a result, unless antitrust law and enforcement are quite sensitive to the problems posed by the patent thicket, they can have the perverse effect of slowing down the commercialization of new discoveries and ultimately retarding innovation, precisely the opposite of the intent of both the patent laws and the antitrust laws.

II. Market Responses to Overlapping Patents

The Economic Theory of Complements

The generic problem inherent in the patent thicket is well understood as a matter of economic theory, at least in its static version. Consider, for example, a company seeking to manufacture a new graphics chip for use in personal computers or video game consoles. (Substitute a biotech firm using patented tools for genetic engineering, or an e-commerce firm using patented business methods, if you would prefer.) Suppose that the company's preferred design for this chip is likely to infringe on a number of patents; the process manufacturing methods used to actually produce the chip infringe on a number of additional patents. In order to produce the chip as designed, the company needs to obtain licenses from a number, call it N, of separate rights holders.

This situation is precisely the classic complements problem originally studied by Cournot in 1838. Cournot considered the problem faced by a manufacturer of brass who had to purchase two key inputs, copper and zinc, each controlled by a monopolist.[5] As Cournot demonstrated, the resulting price of brass was *higher* than would arise if a single firm controlled trade in both copper and zinc, and sold these inputs to a competitive brass industry (or made the brass itself). Worse yet, the combined profits of the producers were lower as well in the presence of complementary monopolies. So, the sad result of the balkanized rights to copper and zinc was to harm both consumers and producers.[6] The same applies today when multiple companies control blocking patents for a particular product, process, or business method.

How can the inefficiency associated with multiple blocking patents be eliminated? One natural and attractive solution is for the copper and zinc suppliers to join forces and offer their inputs for a single, package price to the brass industry. The two monopolist suppliers will find it in their joint interest to offer a package price that is less than these two components sold for when priced separately. The blocking patent version of this principle is that the rights holders will find it attractive to create a package license or patent pool, or in some situations to simply engage in cross licensing so they can each produce final products themselves.

The appendix offers a short, modern, and more general version of Cournot's theory of complements cast in terms of blocking patents. This basic theory of complements (used in fixed proportions) gives strong support for businesses to adopt, and for competition authorities to welcome, either cross licensees, package licenses, or patent pools to clear such blocking positions. If two patent holders are the only companies realistically capable of manufacturing products that utilize their intellectual property rights, a royalty-free cross license is ideal from the point of view of *ex post* competition. But *any* cross license is superior to a world in which the patents holders fail to cooperate, since neither could proceed with actual production and sale in that world without infringing on the other's patents. Alternatively, if the two patent holders see benefits from enabling many others to make products that utilize their intellectual property rights, a patent pool, under which all the blocking patents are licensed in a coordinated fashion as a package, can be an ideal outcome. The simple theory, which is sketched out in the appendix, suggests that coordinating such licensing can lead to *lower* royalty rates than would independent pricing (licensing) of the two companies' patents.

In other words, without cross licenses or patent pools, there is a tendency for products to bear multiple patent burdens. The buildup of licensing fees can have several unattractive consequences. First, the well-known costs of static monopoly power are magnified: prices are well above marginal costs, causing inefficiently low use of these products. As shown in the appendix, with N rights holders, equilibrium markups are N times the monopoly level. Of course, this is merely a magnified version of the monopoly burden resulting from the patent system itself, but it is well to remember Cournot's lesson that the multiple burdens reduce both consumer welfare and the profits of patentees in comparison with a coordinated licensing approach. Second, these burdens may cause certain products not to be produced at all, if that production is subject to economies of scale. Third (this is a dynamic version of the previous point), the prospect of paying such royalties necessarily reduces the return to new product design and development, and thus can easily be a drag on innovation and commercialization of new technologies.

Heller and Eisenberg (1998) discuss the complements problem in the context of biotechnology patents, making a nice comparison to the classic tragedy of the commons. The well-known tragedy of the commons refers to the fact that a resource can be overused if it is not protected by property rights; fishing grounds and clean water are standard examples. Heller and Eisenberg point out that quite a different problem arises when there are multiple blocking patents; they label this problem the tragedy of the anti-commons. The tragedy of the anti-commons arises when there are multiple gatekeepers, each of whom must grant permission before a resource can be used. With such excessive property rights, the resource is likely to be underused. In the case of patents, innovation is stifled.

The Holdup Problem

As noted above, the complements problem is at its worst when the downstream firms using the various inputs truly require each input to make their products. In the patent context, if a manufacturer finds it relatively easy to design around a given patent, the royalties that the patentee can assert are necessarily limited. So, unless the patent in question is quite broad, one might think that any burden on the manufacturer would be modest, and arguably the very return we wish to provide to the patentee as a reward for innovation.

Unfortunately, this rather romantic view of patents is less and less applicable in our economy, for three reasons. First, even a modest tax is counterproductive if the patent was improperly granted, that is, if the patentee did not truly made a new and useful discovery, or if the patent as granted was too broad, covering some prior art as well as something truly new. Second, the cumulative effect of many small taxes can become quite large; there are sound reasons to believe that the static deadweight loss associated with these royalties is increasing and convex in the tax rate, at least over some range of royalties. The danger of paying royalties to multiple patent owners is hardly a theoretical curiosity in industries such as semiconductors in which many thousands of patents are issued each year and manufacturers can potentially infringe on hundreds of patents with a single product.

Third, and most important, is timing. Suppose that our representative manufacturer could, with ease, invent around a given patent, if that manufacturer were aware of the patent and afforded sufficient lead time. Clearly, in this case the patented technology contributes little if anything to the final product, and any reasonable royalty would be modest at best. But, oh, how the situation changes if the manufacturer has already designed its product and placed it into large scale production before the patent issues. In this case, even though the timing is strongly suggestive that the manufacturer did not in fact rely on the patented invention for the design of its product, the manufacturer is in a far weaker negotiating position. The patentee can credibly seek far greater royalties, very likely backed up with the threat of shutting down the manufacturer if the Court indeed finds the patent valid and infringed and grants injunctive relief. The manufacturer *could* go back and redesign its product, but to do so (a) could well require a major redesign effort and/or cause a significant disruption to production, (b) would still leave potential liability for any products sold after the patent issued before the redesigned products are available for sale, and (c) could present compatibility problems with other products or between different versions of this product. In other words, for all of these reasons, the manufacturer is highly susceptible to holdup by the patentee. I submit that this holdup problem is very real today, and that both patent and antitrust policymakers should regard holdup as a problem of first order significance in the years ahead.

The holdup problem is worst in industries where hundreds if not thousands of patents, some already issued, others pending, can potentially read on a given product. In these industries, the danger that a manufacturer will step on a land mine is all too real. The result will be that some companies avoid the mine field altogether, that is, refrain from introducing certain products for fear of holdup. Other companies will lose their corporate legs, that is, will be forced to pay royalties on patents that they could easily have invented around at an earlier stage, had they merely been aware that such a patent either existed or was pending. Of course, ultimately the expected value of these royalties must be reflected in the price of final goods.

In short, with multiple overlapping patents, and under a system in which patent applications are secret and patents slow to issue (relative to the speed of new product introduction), we have a volatile mix of two powerful types of transaction costs that can burden innovation: (1) the complements problem, the solution of which requires coordination, perhaps large scale coordination; and (2) the holdup problem, which is quite resistant to solution in the absence of either (a) better information at an earlier stage about patents likely to issue, and/or (b) the ability of interested parties to challenge patents at the PTO before they have issued and are given some presumption of validity by the Courts.

Clearly, these concerns form the basis for a serious discussion about reform of the patent system.[7] However, my intention in this paper is to explore how private companies can best navigate the patent system we currently have, and how our antitrust laws can be enforced in a way that is sensitive to the transaction costs associated with our current patent system. I see relatively little that private companies can do to overcome the holdup problem without reform of the patent system itself. But there is quite a bit they can do to solve the complements problem, which itself is greatly exacerbated by the holdup problem.

Overlapping Patents and Business Strategy in Practice

To solve the complements problem generally, and to cut through the patent thicket specifically, requires coordination among rights holders. Such coordination itself faces two types of obstacles. First, there are inevitably coordination costs that must be overcome. Second, antitrust sensitivities are invariably heightened when companies in the same or related lines of business combine their assets, jointly set fees of any sort, or even talk directly with one another. Because such coordination may involve the elimination of competition, we have a complex interaction between private and public interests. Even as coordination between rights holders is critical, from a public policy perspective we cannot presume that private deals are in the public interest. Antitrust authorities will legitimately want to know whether consumers are helped or harmed by any arrangement; injured parties may seek redress under the antitrust laws or by alleging patent misuse.

Cross Licenses   Cross licenses commonly are negotiated when each of two companies has patents that may read on the other's products or processes. Rather than blocking each other and going to court or ceasing production, the two enter into a cross license. Especially with a royalty free cross license, each firm is then free to compete, both in designing its products without fear of infringement and in pricing its products without the burden of a per unit royalty due to the other. Thus, cross licenses can solve the complements problem, at least among two firms, and thus be highly procompetitive.

A cross license is simply an agreement between two companies that grants each the right to practice the other's patents. Cross licenses may or may not involve fixed fees or running royalties; running royalties can in principle run in one direction or both. Cross licenses may involve various field-of-use restrictions or geographic restrictions. Cross licenses may involve some but not all relevant patents held by either party; carve-outs are not uncommon. And cross licenses, like regular licenses, may be confined to patents issued (or pending) as of the date of the license, or they may include patents to be granted through a certain time in the future.

Patent Pools and Package Licenses   When two or more companies control patents necessary to make a given product, and when at least some actual or potential manufacturers may not themselves hold any such patents, a patent pool or a package license can be the natural solution to the complements problem. Under a patent pool, an entire group of patents is licensed in a package, either by one of the patent holders or by a new entity established for this purpose, usually to anyone willing to pay the associated royalties. Under a package license, two or more patent holders agree to the terms on which they will jointly license their complementary patents and divide up the proceeds. A nice example of a patent pool is the Manufacturers Aircraft Association formed in 1917 to license a number of patents necessary for the production of

airplanes, patents controlled by The Wright-Martin Aircraft Corporation, the Curtiss Aeroplane & Motor Corporation, and others.[8] I discuss
below some more recent patent pools that have been used to help establish compatibility standards.

Cooperative Standard Setting   The need to solve the complements problem tends to be especially great in the context of standard setting. For
example, when the International Telecommunications Union (ITU) establishes a new standard for fax transmissions or modem protocols, the
participants are loath to agree to a standard that can be controlled by
any single firm through its patents. Thus, standard setting organizations like the ITU or the American National Standards Institute (ANSI)
typically require that participants agree to license all patents essential
to compliance with any standard on "fair, reasonable, and nondiscriminatory" terms. Rules such as this are explicitly intended to reduce or
eliminate any holdup problems. However, it is well to note that many
standard setting organizations are wary of sanctioning any specific
agreement regarding the magnitude of licensing terms for fear of antitrust liability, as such agreements might be construed as price fixing.
Perversely, by leaving the precise licensing terms vague, this caution
can in fact lead to ex post holdup by particular rights holders, contrary
both to the goal of enabling innovation and to consumers' interests.

The case in which multiple firms control patents essential to a standard fits well with the formal economic analysis described above. In essence, any manufacturer seeking to produce a compliant product must
obtain a license from each rights holder to avoid facing an infringement action. Inventing around is typically impractical, as it would preclude the manufacturer from claiming that its products are compliant
and thus assuring consumers that they are fully compatible with the
prevailing standard. Thus, standard setting very often has especially
strong elements of both the complements problem and the holdup
problem.

Settlements of Patent Disputes   Cross licenses (or simply licenses) are a
common way in which companies resolve patent disputes. But other
forms of settlement arise, two of which I touch on below. First, I discuss
acquisitions, in which one firm simply acquires the other, thereby re-
solving the dispute and assembling the various intellectual property
rights within a single company. Second, I comment on cash payments
in exchange for exit, a strategy whereby one company pays the other
company to exit the market, and thus to drop its challenge to the first

company's patent. In each of these cases, legitimate questions arise as to whether any particular private agreement truly is in the public interest.

Antitrust Limits

As I have indicated, many of the business solutions to the complements problem and the holdup problem raise antitrust issues. Quite generally, agreements among companies that either do compete, or might compete, directly with each other raise antitrust warning flags. For each business form, I consider below its antitrust treatment.

Generally speaking, one can imagine two rather different approaches that antitrust might take to firms' efforts to coordinate to solve the complements problem. One approach is to ask whether the agreement in question leads to more competition than would occur *without* that agreement. This is the approach advocated in the Department of Justice and Federal Trade Commission *Antitrust Guidelines for the Licensing of Intellectual Property*, which state in §3.1 that:

However, antitrust concerns may arise when a licensing arrangement harms competition among entities that would have been actual or likely potential competitors in a relevant market in the absence of the license (entities in a "horizontal relationship").

Another quite different approach would be to ask whether the agreement in question is the most competitive agreement possible. Put differently, one could ask whether a given agreement is the least restrictive alternative that is workable in the sense of solving the legitimate business problem faced, such as unblocking patent positions. Clearly, this latter standard, which does not reflect current antitrust enforcement policy according to the Guidelines, would be far tougher on all forms of cooperation among patent and copyright holders.

III.   Cross Licensing

Cross Licenses and Design Freedom

Cross licenses are the preferred means by which large companies clear blocking patent positions amongst themselves. Based in part on work I have done on behalf of Intel, I can report that broad cross licenses are the norm in markets for the design and manufacture of microprocessors.[9] For example, Intel has entered into a number of broad cross licenses with other major industry participants, such as IBM, under which most of each company's vast patent portfolio is licensed to the other. Furthermore, the companies generally agree to grant licenses to each other for patents that will be issued several years into the future, typically for the lifetime of the cross licensing agreement. Often, these cross licenses involve no running royalties, although they may involve balancing payments at the outset to reflect differences in the strength of the two companies' patent portfolios as reflected in a patent pageant, and/or the vulnerability of each to an infringement action by the other. For example, Hewlett-Packard and Xerox recently announced a cross license that settled their outstanding patent disputes.

From the perspective of competition policy, cross licenses of this sort are quite attractive. The traditional concern with cross licenses among competitors is that running royalties will be used as a device to elevate prices and effect a cartel; see Katz and Shapiro 1985. Clearly, such concerns do not apply to licenses that involve small or no running royalties, but rather have fixed up-front payments. Another concern is that the granting of licenses to *future* patents will reduce each company's incentive to innovate because its rival will be able to imitate its improvements.[10] While correct in theory, it is clear, at least in the case of semiconductors and no doubt more widely, that this concern is dwarfed by the benefits arising when each firm enjoys enhanced design freedom by virtue of its access to the other firm's patent portfolio. There is little doubt that these broad cross licenses permit the more efficient use of engineers (arguably the resource that governs the rate of innovation in the semiconductor industry), better products, and faster product design cycles. In other words, when IBM and Intel sign a forward looking cross license, each is enabled to innovate more quickly and more effectively without fear that the other will hold it up by asserting a patent that it has unintentionally infringed. And neither firm is really all that worried that the other will actually *copy* its products, just because the other has a license to most of its patents. Of course, the impressive rate of innovation in the semiconductor industry in the presence of a web of such cross licenses offers direct empirical support for the view that these cross licenses promote rather than stifle innovation.

Intel's Policy of "IP for IP"

Despite all of these benefits, the Federal Trade Commission attacked Intel's cross licensing practices in 1998.[11] One key episode behind the

FTC's complaint involved Intel's conduct when faced with a lawsuit by Intergraph, a workstation manufacturer, asserting that Intel's microprocessors infringed on certain patents held by Intergraph. Of course, lawsuits like Intergraph's are a necessary part of the threat point behind any cross-licensing negotiation: if one party is not happy with the terms offered by the other, it always has the option of initiating patent litigation. In response to Intergraph's infringement action against Intel, Intel withdrew its own intellectual property from Intergraph by suing Intergraph for infringement of Intel's patents and by withdrawing the supply of Intel trade secrets to Intergraph, trade secrets that Intergraph valued highly for the purposes of designing systems built on Intel chips.

Evidently viewing Intel's conduct as unfair, the FTC attempted to fashion an antitrust case against Intel based on this conduct, along with a similar response by Intel to a lawsuit initiated by Digital Equipment Corporation.[12] The FTC action against Intel sharply exposed the fact that the FTC and Intel had fundamentally different views about the impact of the conduct at issue. The FTC saw Intel as using its existing monopoly power to fortify its position by lowering its royalty costs per chip and potentially offering superior products by incorporating technologies patented by others. Intel viewed itself as engaging in a defensive exercise which was a necessary aspect of cross licensing, namely trading intellectual property for intellectual property (IP for IP) and withdrawing its own intellectual property when faced with a frontal assault on its core product line in the form of an infringement action seeking injunctive relief. Intel, well aware of what a juicy target it posed, believed it had every right to protect itself from holdup, and certainly no duty to give special treatment in the form of Intel trade secrets and advance product samples to a company attempting to hold it up.

The problem for the FTC was that the conduct at issue, especially with respect to Intergraph, was directed at a *customer* of Intel's, not a competitor. Brushing aside concerns about holdup, and playing down the important role of cross licenses in the semiconductor industry, the FTC found no "business justification" for Intel's conduct, and thus was prepared to presume that the conduct was anticompetitive without actually studying the impact of the conduct on Intel's competitors. In fact, Intel's true rivals in microprocessor design and manufacturing (such as AMD, Motorola, Sun, or IBM) were either not subject to the conduct at issue (since they were not Intel customers at all and thus not recipients of the Intel trade secrets at issue), or had ongoing cross licenses with Intel under which the litigation triggering these episodes would simply not occur in the first place.

Fortunately, a compromise was reached and a settlement agreed to between the FTC and Intel.[13] In essence, Intel agreed not to withdraw product information needed by its customers to build systems based on soon-to-be-released Intel chips. (Presumably, this promise provides some benefit to Intel by assuring its customers that *they* will not be held up once they are relying on Intel for their new systems.) But Intel is not obligated to continue to provide trade secrets on products farther out on their roadmap (i.e., products that will not be introduced for a year or two) to customers suing Intel, and Intel was not obligated to provide *any* trade secrets to a company suing Intel and seeking a court injunction to shut down Intel's microprocessor business.

The Intel situation also exposes the interplay between government enforcement of the antitrust laws and private antitrust actions. Even while the FTC was investigating Intel, bringing a complaint against Intel, and ultimately settling with Intel, Intergraph was engaged in its own antitrust and patent battle with Intel. Intergraph won a resounding victory in the first round of that battle, in which the District Court judge in Alabama issued a searing anti-Intel opinion ruling, among other things, that Intel's microprocessors and associated trade secrets were "essential facilities" under antitrust laws, thus imposing a duty on Intel to sell its microprocessors to Intergraph and to make its trade secrets available to Intergraph, Intergraph's lawsuit against Intel notwithstanding. This opinion was based on strands of antitrust law that require dominant companies to deal with their rivals, especially if the dominant firm has established an ongoing course of dealing with rivals in the past.[14]

Ultimately, however, Intel was vindicated. The District Court judge later ruled that Intel was not in fact infringing on Intergraph's patents. And, most significantly, the Court of Appeals for the Federal Circuit vacated the District Court's antitrust and essential facility opinion.[15] In a strongly worded and sweeping opinion, the appeals court ruled that Intel's conduct did not violate the antitrust laws because it was not directed at a competitor and indeed could have no adverse impact on competition in the market where Intel was alleged to have monopoly power, namely the market for microprocessors, in which Intergraph did not compete. The FTC's efforts to fashion an antitrust case out of Intel's conduct look even more dubious now in the light of this subsequent decision by the Court of Appeals.

The Intel episode is closely related to another ongoing debate regarding the intersection between intellectual property rights and antitrust law: can a company violate the antitrust laws simply by refusing to license its patents, or by refusing to sell patented items, to its rivals? Most commentators have said for some time that a refusal to license patents cannot in and of itself constitute an antitrust violation. However, the Supreme Court has signaled that unilateral refusals to sell can indeed constitute antitrust violations, especially if a company has established an ongoing course of dealing with its rivals.[16] The precise conditions under which a refusal to license a patent (or to sell patented items) could constitute an antitrust violation has remained unclear. Most observers were stunned when the Ninth Circuit Court of Appeals ruled in 1997 that Kodak was liable for refusing to sell patented spare parts for its machines to independent service organizations seeking to compete against Kodak in the business of servicing Kodak copiers and micrographics equipment. As the Court acknowledged, this was the first time a unilateral refusal to sell a patented item had been judged to be an antitrust violation.[17] Just recently, the Court of Appeals for the Federal Circuit came to a very different conclusion, ruling that a company's unilateral decision not to license a patent (or sell a patented item) could *never* in and of itself constitute an antitrust violation.[18] Hopefully, the Supreme Court will resolve this significant split among the Circuit Courts and clarify that unilateral refusals to license patents are immune from antitrust challenge.

Intel's practices, and those of other firms who require grantbacks of relevant patents in exchange for a license to key enabling patents, copyrights, or trade secrets, raises further interesting questions about the role of self help in the digital economy.[19] One view of such business strategies cum legal regimes is that they are a welcome effort by leading firms to establish a type of litigation-free zone likely to favor innovation and get around some of the current difficulties with our patent system and the patent thicket it causes. A less favorable view is that these arrangements represent efforts by powerful firms to establish private legal regimes that favor themselves and make it more difficult for upstarts to challenge the dominance of current market leaders. Is a cross licensing policy of IP for IP a beneficial way to cut through the patent thicket, or a strong-arm tactic by a dominant firm that enjoys powerful patent rights and seeks access to others' intellectual property in exchange?

IV.   Patent Pools

A patent pool involves a single entity (either a new entity or one of the original patent holders) that licenses the patents of two or more companies to third parties as a package. In many respects, a patent pool (much like a package license) is the purest solution to the complements problem described above and analyzed in the appendix. Indeed, licensees may well welcome such a pool, both for the convenience of one-stop shopping and because a subset of the required patents may be of little or no value by themselves. Thus, from the licensee's perspective, licensing the entire package is simpler and avoids the danger of paying for some patent rights that turn out to be useless without other complementary rights.

Essential Patents vs. Rival Patents

The Department of Justice (DOJ) has clearly articulated its policy toward patent pools/package licensing in a trio of business review letters regarding an MPEG patent pool and two DVD patent pools. The essence of this approach, which precisely mirrors the economic principles articulated above, is that inclusion of truly *complementary* patents in a patent pool is desirable and procompetitive, but assembly of *substitute* or rival patents in a pool can eliminate competition and lead to elevated license fees. But differently, the key distinction in forming a patent pool is that between *blocking* or *essential* patents, which properly belong in the pool, and *substitute* or *rival* patents, which may need to remain separate.

In the MPEG case,[20] the Department approved the creation of a pool of patents necessary to enable manufacturers to meet the MPEG-2 video compression technology. This pool, encompassing patents from Fujitsu, General Instrument, Lucent, Matsushita, Mitsubishi, Philips, Scientific-Atlanta, Sony, and Columbia University, permits one-stop shopping for makers of televisions, digital video disks and players, and telecommunications equipment as well as cable, satellite, and broadcast television services. To support their formation of a patent pool, these nine patent holders conducted an extensive search to identify all patents essential to the MPEG-2 standard and include them in the pool. The licensing agent for the pool, MPEG LA, will employ an independent patent expert to determine whether a patent in the pool is in fact essential, and whether other patents as well are essential and thus suitable for inclusion in the pool. As stated by the Department, "the use of the independent-expert mechanism will help ensure that the portfolio will contain only patents that are truly essential to the MPEG-2 standard, weeding out patents that are competitive alternatives to each other."

In the first Digital Versatile Disk (DVD) case,[21] the Department approved a proposal by Philips, Sony, and Pioneer to jointly license patents necessary to make discs and players that comply with the DVD-Video and DVD-ROM standards. Again, only essential patents are to be included in the joint licensing program. As with the earlier CD licensing program of Sony and Philips, licenses will be offered by Philips, in this case on behalf of all three firms. Again, an independent patent expert will be employed to ensure that the license only conveys the rights to essential patents. As stated by the Department, "the expert will help ensure that the patent pool does not combine patents that would otherwise be competing with each other." The Department subsequently approved a second joint licensing scheme relating to the DVD-Video and DVD-ROM standards,[22] this one including patents held by Toshiba (the licensing entity), Hitachi, Matsushita, Mitsubishi, Time Warner, and Victor Company of Japan. Note that the effect of these two patent pools appears to be to reduce but not eliminate the complements problem, since there remain two separate pools, not just one: two-stop shopping, it would appear.

A Patent Pool Created to Resolve Claims of Blocking Patents

In contrast to the Department of Justice's approval of these three patent pools, the Federal Trade Commission in March 1998 challenged a patent pool formed by Summit Technology, Inc. and VisX, Inc., two firms that manufacture and market lasers to perform a new, and increasingly popular, vision correcting eye surgery, photorefractive keratectomy.[23] According to the FTC: "Instead of competing with each other, the firms placed their competing patents in a patent pool and share the proceeds each and every time a Summit or VISX laser is used." The FTC was ostensibly following the same principles employed by the Justice Department, namely to permit the assembly of complementary or essential patents, but not rival patents, into a pool. According to the FTC, the two companies agreed not to license their patents independently.

However, the companies in this case argued vigorously that they did indeed have mutually blocking patents, making their pool, Pillar Point Partners, procompetitive. In August 1998 the two companies settled with the FTC and agreed to lift any restrictions on each other regarding the licensing of their patents; ultimately, their patent pool was dissolved.[24]

The Summit and VisX case raises a number of very interesting and tricky issues regarding patent pools and joint licensing programs in general. First, if two companies reasonably believed that their patents blocked each other at the time they formed the pool, was that sufficient to justify the formation of a pool? How hard are they required to look into the validity of each other's claims before agreeing to pool their patents? Second, if each firm believed it could, at considerable expense, delay, and risk, invent around the other's patents, should the two firms be prohibited from forming a pool and rather forced to attempt to invent around each other's patents, under the view that consumers *might* thereby enjoy the benefits of direct competition (although the product might be delayed, or never introduced, in the absence of the pool)? Third, is there competitive harm in placing some potentially rival patents into the pool, assuming that each party in fact controls valid blocking patents, making *some* type of pool procompetitive? Fourth, can the pool be attacked on antitrust grounds based on the argument that a less restrictive alternative, namely a cross license, would have achieved the same legitimate purposes and created additional *ex post* competition? If so, does it matter in this assessment if the two companies agree that the pool will license their patents to third parties, something that a cross license would not permit, unless it contained rather unusual sublicensing rights?

V. Cooperative Standard Setting

Blocking patents are especially common in the context of standard setting: once a standard is picked, any patents (or copyrights) necessary to comply with that standard become truly essential. If the standard becomes popular, each such patent can confer significant market power on its owner, and the standard itself is subject to holdup if these patent holders are not somehow obligated to license their patents on reasonable terms. As noted above, for precisely this reason, standard setting bodies require participants to license any essential patents on reasonable terms as a quid pro quo before adopting any standards.[25]

Fortunately, antitrust concerns have not prevented a great many cooperative standard setting efforts from proceeding forward. Some participants go so far as to say that much of the innovation taking place now in the telecommunications, Internet, and computer areas is standards based. Indeed, even the fiercest enemies often team up in the software industry to promote new standards. Back in 1997, Microsoft and Netscape, two companies hardly known as cozy partners, agreed to include compatible versions of Virtual Reality Modeling Language (developed by Silicon Graphics) in their browsers. This agreement was expected to make it far easier for consumers to view 3D images on the Web. Earlier, Microsoft agreed to support the Open Profiling Standard, which permits users of personal computers to control what personal information is disclosed to a particular web site, and which had previously been advanced by Netscape, along with Firefly Network, Inc. and Verisign Inc.

But neither is cooperative standard setting immune from antitrust scrutiny. In the consumer electronics area, for example, the Justice Department investigated Sony, Philips, and others regarding the establishment of the CD standard in the 1980s. Cooperative efforts to set optical disc standards have also been challenged in private antitrust cases, on the theory that agreements to adhere to a standard are an unreasonable restraint of trade:

[d]efendants have agreed, combined, and conspired to eliminate competition . . . by agreeing not to compete in the design of formats for compact discs and compact disc players, and by instead agreeing to establish, and establishing, a common format and design . . .[26]

Does cooperation lead to efficient standardization, increased competition, and additional consumer benefits? Or is cooperative standard setting a means for firms collectively to stifle competition, to the detriment of consumers and firms not included in the standard setting group? Answering these questions and evaluating the limits that should be placed on cooperative standard setting efforts require an analysis of the competitive effects of such cooperation in comparison with some reasonable but-for world. Inevitably, an antitrust analysis of cooperative standard setting involves an assessment of how the market would likely evolve *without* the cooperation. One possibility is that multiple, incompatible products would prevail in the market, if not for the cooperation. Another possibility is that the market would eventually tip to a single product, even without cooperation. Even in this latter case, an initial industrywide standard can have significant efficiency and welfare consequences, for three reasons: (1) cooperation may lock in a different product design than would emerge from competition; (2) cooperation may eliminate a standards war waged prior to tipping; and (3) cooperation is likely to enable multiple firms to supply the industry standard product, whereas a standards war may lead to a single, proprietary product.

The Costs and Benefits of Compatibility and Standards

There are significant benefits associated with achieving compatibility. These include:

- successful launching of a bandwagon or network,
- greater realization of network effects,
- protecting buyers from stranding, and
- enabling competition within an open standard.

Likewise, standardization and compatibility can impose very real costs on consumers:

- constraints on variety and innovation,
- loss of ex ante competition to win the market, and
- proprietary control over a closed standard.

Legal Treatment of Cooperative Standard Setting

I now look more closely at the intellectual property issues that arise specifically in the context of standard setting, where the participants typically agree to license their patents on fair, reasonable, and nondiscriminatory terms.

Firms are sometimes accused of hiding intellectual property rights until after the proprietary technology has been embedded in a formal standard. I view this issue primarily as one of contract law. Standard setting groups typically have provisions in their charters compelling participants either to reveal all relevant intellectual property rights or to commit to licensing any intellectual property rights embedded in the standard on reasonable terms.[27] Clearly, these rules help control the holdup problem. In some cases, however, the precise requirements imposed by a standard setting group may be unclear. In these circumstances, if the standard affects nonparticipants, including consumers, there is a public interest in clarifying the duties imposed on participants in a fashion that promotes rather than stifles competition.

The question of whether firms should be allowed, or even encouraged, to set standards cooperatively is part of the broader issue of collaboration among competitors, a storied area within antitrust law. Most of the case law deals with quality and performance standards rather than compatibility standards.[28] Existing cases also have tended to focus on the standard setting process itself, rather than the outcomes of cooperative standard setting.

Antitrust liability has been found for participants in a standard setting process who abuse that process to exclude competitors from the market. One leading case is *Allied Tube & Conduit Corp. v. Indian Head, Inc.*, 486 U.S. 492 (1988), in which the Supreme Court affirmed a jury verdict against a group of manufacturers of steel conduit for electrical cable. These manufacturers conspired to block an amendment of the National Electric Code that would have permitted the use of plastic conduit. They achieved this by packing the annual meeting of the National Fire Protection Association, whose model code is widely adopted by state and local governments. The other leading case is *American Society of Mechanical Engineers v. Hydrolevel Corp.*, 456 U.S. 556 (1982), in which the Supreme Court affirmed an antitrust judgment against a trade association. In this case, the chairman of an association subcommittee offered an unofficial ruling that the plaintiff's product was unsafe, and this ruling was used by the plaintiff's rival (who enjoyed representation on the subcommittee) to discourage customers from buying the plaintiff's product.

Antitrust risks associated with excluding a rival from the market appear to be less of a problem for an open standard, but could arise if the companies promoting the standard block others from adhering to the standard or seek royalties from outsiders. The DOJ business review letters regarding the MPEG-2, DVD-Video, and DVD-ROM standards are excellent illustrations of how the enforcement agencies can successfully handle intellectual property in the standard setting context.

As the Supreme Court has noted, "Agreement on a product standard is, after all, implicitly an agreement not to manufacture, distribute, or purchase certain types of products."[29] To date, this type of reasoning has not been used to impose per se liability on software standard setting activities. Indeed, I know of no successful antitrust challenges to cooperation to set compatibility standards. The closest case of which I am aware is *Addamax Corporation v. Open Software Foundation, Inc.*,

888 F. Supp. 274 (1995). In *Addamax*, the District Court refused to grant summary judgment on behalf of the Open Software Foundation, an industry consortium formed to develop a platform-independent version of the UNIX operating system. OSF conducted a bidding to select a supplier of security software. After failing to be selected, Addamax brought antitrust claims against OSF, Hewlett-Packard, and Digital Equipment Corporation, asserting that OSF had chosen the winner not based on the merits but to favor specific companies and technologies. The Addamax case looks problematic, inasmuch as the primary purpose of OSF was to permit its members to team up to offer stronger competition against the leading UNIX vendors, Sun Microsystems and AT&T, and there was no evidence suggesting that OSF's failure to pick Addamax was based on its members desire to control the market in which Addamax itself operated.

Ultimately, the antitrust risks faced by companies that are trying to set compatibility standards appear to be relatively minor as long as the scope of the agreement truly is limited to standard setting and steers clear of distribution, marketing, and pricing. While the law has typically looked for integration and risk-sharing among collaborators in order to classify cooperation as a joint venture and escape per se condemnation, these are not very helpful screens for standard setting activities. The essence of cooperative standard setting is not the sharing of risks associated with specific investments, or the integration of operations, but rather the contribution of complementary intellectual property rights and the expression of unified support to ignite positive feedback for a new technology.

The limits imposed by public policy in the area of compatibility standards remain unclear. The most specific statement by the antitrust enforcement agencies can be found in a recent FTC Staff Report.[30] The Staff Report recognized a need for clarification in this area:

the time has come for a significant effort to rationalize, simplify, and articulate in one document the antitrust standards that federal enforcers will apply in assessing collaborations among competitors. This effort should be directed at drafting and promulgating "competitor collaboration guidelines" that would be applicable to a wide variety of industry settings and flexible enough to apply sensibly as industries continue rapidly to innovate and evolve.[31]

Since that call for action, the FTC has conducted Joint Venture Hearings, and the Commission and the Antitrust Division issued in

April 2000 new "Antitrust Guidelines for Collaborations Among Competitors" (available at either Agency's web site).

Hidden Patents and Holdup in Standard Setting

A number of disputes have surfaced recently that illustrate the thorny problems associated with hidden patent rights that were later exerted against established standards.[32]

Dell Computer and the VESA VL-Bus Standard   The leading U.S. example of this type of antitrust action is the FTC's consent agreement with Dell Computer Corporation, announced in November 1995. Although the case involved computer hardware, it is important for the software community as well. The assertion was that Dell threatened to exercise undisclosed patent rights against computer companies adopting the VL-bus standard, a mechanism to transfer data instructions between the computer's CPU and its peripherals such as the hard disk drive or the display screen. The VL-bus was used in 486 chips, but it has now been supplanted by the PCI bus. According to the FTC.

During the standard-setting process, VESA [Video Electronics Standard Association] asked its members to certify whether they had any patents, trademarks, or copyrights that conflicted with the proposed VL-bus standard; Dell certified that it had no such intellectual property rights. After VESA adopted the standard—based in part, on Dell's certification—Dell sought to enforce its patent against firms planning to follow the standard.[33]

There are two controversial issues surrounding this consent decree: (a) the FTC did not assert that Dell acquired market power, and indeed the VL-bus never was successful; and (b) the FTC did not assert that Dell *intentionally* misled VESA. My analysis suggests that anticompetitive harm is unlikely to arise in the absence of significant market power and that the competitive effects are not dependent on Dell's intentions.

Motorola and the ITU V.34 Modem Standard   Another good example of how competition can be affected when standard setting organizations impose ambiguous duties on participants is the case of Motorola and the V.34 modem standard adopted by the International Telecommunications Union. Motorola agreed to license its patents essential to the standard case to all comers on "fair, reasonable, and nondiscriminatory terms."[34] Once the standard was in place, Motorola then made offers that some industry participants did not regard as meeting this obligation. Litigation ensued between Rockwell and Motorola, in part over the question of whether reasonable terms should mean: (a) the terms that Motorola could have obtained ex ante, in competition with other technology that could have been placed in the standard; or (b) the terms that Motorola could extract ex post, given that the standard is set and Motorola's patents are essential to that standard.

These issues are best dealt with by the standard setting bodies, or standard setting participants, either by making more explicit the duties imposed on participants, or by encouraging ex ante competition among different holders of intellectual property rights to get their property into the standard. Unfortunately, antitrust concerns have led at least some of these bodies to steer clear of such ex ante competition, on the grounds that their job is merely to set technical standards, not to get involved in prices, including the terms on which intellectual property will be made available to other participants. The ironic result has been to embolden some companies to seek substantial royalties after participating in formal standard setting activities.

VI. Settlements of Patent Disputes

Cross licenses and patent pools can be ways to settle intellectual property disputes. For example, the Summit and VisX patent pool discussed above, Pillar Point Partners, was essentially a settlement of a patent dispute between Summit and VisX.

Generally speaking, antitrust authorities have legitimate concerns that parties will settle their intellectual property disputes in ways that stifle competition. As a matter of economic theory, there is no reason to expect the two parties' collective interests in settlement, and especially in the *form* of any settlement they adopt, to coincide with the public interest, which includes consumer interests. So, while the law surely welcomes the settlement of disputes generally, and does not seek to force parties to litigate to the death, some settlements can be anticompetitive. Based on this general view, Assistant Attorney General for Antitrust Joel Klein recently suggested (see Klein 1997) that parties notify the Justice Department of certain settlements that they enter into, much as parties are required to notify the Justice Department and the FTC in advance of their intention to merge.

Firms are quite creative in crafting settlements of intellectual property disputes, and by no means restrict their attention to cross licenses and patent pools. For example, one tried and true method of settling a dispute is for the companies involved simply to merge. However, the antitrust authorities are well aware that such mergers can themselves eliminate competition, and they will view such mergers with skepticism if there is a good chance that the two parties will in fact be capable of competing against each other, their patent claims notwithstanding. A good example of such a merger that was modified in response to FTC concerns was the proposed merger of Boston Scientific and CVIS in the area of imaging catheters.[35] An interesting twist in such cases is that the parties' posturing in court, where they each have an incentive to assert that they are not infringing on the other's patents, provides direct ammunition to the FTC or DOJ to assert that the two companies could indeed compete independently if not for the merger.

A second method that companies can use to settle a patent dispute is for one company to simply pay the other company to drop its claims and exit the market. Such agreements raise obvious antitrust concerns, because an incumbent firm may be willing to pay handsomely to eliminate a potential competitor and avoid the risk of having its patent challenged, especially if no equally effective challenger is likely to arrive on the scene any time soon. The losers in such deals can easily be subsequent would-be entrants (if the patent were struck down) or consumers (who would benefit from a finding that the patent at issue is invalid or not infringed). Put differently, a settlement can generate negative externalities, either to other firms or to consumers, and thus there is a legitimate role of the Courts and the antitrust enforcement agencies to oversee such settlements.

One class of settlements that are suspicious on their face is that involving agreements between incumbent manufacturers of branded pharmaceuticals and would-be rivals who seek to offer generic competition by challenging the validity of the patents underlying the branded product's dominant position. It has been reported recently that the FTC is considering challenging several such settlements.[36] These cases have an interesting twist resulting from the fact that certain generic manufacturers can gain preferential rights to enter the market before others are permitted to do so. As a result, the branded manufacturer may be able to stall competition by entering into a suitable agreement with the uniquely-placed generic manufacturer, knowing that subsequent rivals will face some delay. In order to identify and prevent any anticompetitive agreements of this nature, the FTC has asked that the FDA require companies to notify the FDA of any such settlements and make that information available to the FTC for its review.

VII. Conclusions

Our current patent system is causing a potentially dangerous situation in several fields, including biotechnology, semiconductors, computer software, and e-commerce, in which a would-be entrepreneur or innovator may face a barrage of infringement actions that it must overcome to bring its product or service to market. In other words, we are in danger of creating significant transaction costs for those seeking to commercialize new technology based on multiple patents, overlapping rights, and holdup problems. Under these circumstances, it is fair to ask whether the pendulum has swung too far in the direction of strong patent rights, ranging from the standards used at the Patent and Trademark Office for approving patent applications, to the secrecy of such applications, to the presumption afforded by the courts to patent validity, to the right of patent holders to seek injunctive relief by insisting that infringing firms cease production of the offending products.

Under these circumstances, we can ill afford to further raise transaction costs by making it difficult for patentees possessing complementary and potentially blocking patents to coordinate to engage in cross licensing, package licensing, or to form patent pools. Yet antitrust law can potentially play such a counterproductive role, especially since antitrust jurisprudence starts with a hostility toward cooperation among horizontal rivals.

So far, the Department of Justice has displayed a keen understanding of the need for those holding complementary rights to coordinate in the licensing of those rights, but the Federal Trade Commission has exhibited less restraint, and arguably is making it more difficult for firms to engage in cross licenses, to offer package licenses, or to form procompetitive patent pools. Many of these issues are likely to be extremely important in the near future, especially with the rise of standard setting as an essential part of the process by which new technologies are commercialized.

Notes

Prepared for presentation at "Innovation Policy and the Economy," National Bureau of Economic Research, Adam Jaffe, Joshua Lerner, and Scott Stern, organizers, April 11, 2000, Washington DC. Comments are welcomed; please direct any comments to shapiro@haas.berkeley.edu.

1. For example, in 1995 Joseph Stiglitz, then Chairman of the Council of Economic Advisors, stated at the opening of the Federal Trade Commission's hearings on Competition Policy in the New High-Tech, Global Marketplace, that "some people jump . . . to the conclusion that the broader the patent rights are, the better it is for innovation, and that isn't always correct, because we have an innovation system in which one innovation builds on another. If you get monopoly rights down at the bottom, you may stifle competition that uses those patents later on and so . . . the breadth and utilization of patent rights can be used not only to stifle competition, but also have adverse effects in the long run on innovation." See FTC Staff Report, p. 6.

2. See http://www.amazon.com/exec/obidos/subst/misc/patents.html/103-4266077-5496631.

3. Nearly 5,000 patents were granted in the U.S. in a recent single year, 1998, relating to microprocessors alone, not to mention semiconductors more broadly.

4. See, for example, Kortum and Lerner 1998, Cohen et al. 2000, and Hall and Ham 1999.

5. For a brief description of Cournot's original work on complements, and modern extensions, see Shapiro 1989, p. 339.

6. Cournot assumed that the two inputs, copper and zinc, were required in certain fixed proportions for the production of brass. If one input can be substituted for the other, they have properties of *substitutes* as well as *complements,* in which case competition between the two input owners can go far to solving the problem posed here. Throughout this paper, I am assuming that the company in question requires rights to practice each of several patents, and that one patent license cannot substitute for another. Clearly, to the extent that a manufacturer, for example, can rely on multiple designs or production processes covered by separate patents with separate owners, the patent thicket is far less of a problem. But even in this relatively friendly setting, extra difficulties can still be raised by the holdup problem, discussed below.

7. For a thoughtful discussion of possible reforms at the Patent and Trademark Office, see Merges 1999.

8. See Klein 1997 for a further description of this pool and how it operated. In this case, the Assistant Secretary of the Navy, Franklin D. Roosevelt, had to lean on the industry to form a pool and help enable wartime production of aircraft.

9. See Hall and Ham 1999 and Grindley and Teece 1997 for additional studies of licensing practices in the semiconductor industry.

10. This concern about discouraging innovation also arises with respect to grantbacks, under which one company agrees to license its future patents in exchange for rights to use an existing patent held by another company. See Gilbert and Shapiro 1997 for a further discussion of grantbacks.

11. In the matter of *Intel Corporation,* Docket No. 9288, Complaint filed June 8, 1998. The

Complaint is available at http://www.ftc.gov/os/1998/9806/intelfin.cmp.htm. I was retained by Intel to work on this matter.

12. For one well-informed articulation of the theory underlying the FTC's position, see Baker 1999.

13. For more information on the settlement between the FTC and Intel, see http://www.ftc.gov/os/1999/9903/d09288intelagreement.htm.

14. The key recent Supreme Court case here is *Aspen Skiing Company v. Aspen Highlands Skiing Corp.,* 472 U.S. 585 (1985), although the essential facilities doctrine goes back to the case of *U.S. v. Terminal Railroad Association of St. Louis,* 224 U.S. 383 (1912).

15. *Intergraph Corporation v. Intel Corporation,* United States Court of Appeals for the Federal Circuit, 98-1308, Decided November 5, 1999, Judge Newman writing the opinion for the Court.

16. The classic cites are *Otter Tail Power Co. v. U.S.,* 410 U.S. 366 (1973) (duty to sell wholesale electric power to a retail competitor) and *Aspen Skiing Company v. Aspen Highlands Skiing Corp.,* 472 U.S. 585 (1985), (duty to continue to offer a joint lift ticket with a rival ski slope).

17. The Court set up a tortured standard under which a company's decision to refuse to license its patent was "presumptively valid," but could be overcome by evidence that the company's *intent* was anticompetitive. Of course, asserting intellectual property rights against a would-be rival is typically anticompetitive in the sense of trying to eliminate a competitor (or at least earn royalties from the competitor, which add to the competitor's costs), so this test is not in fact workable. Amazingly, the Court said that Kodak would be justified in refusing to sell patented parts if its intent was to earn a return on its R&D investment required to design and manufacture those parts, but not if its intent was to eliminate competitors who rely on those very patented parts. I testified on behalf of Kodak in this case.

18. United States Court of Appeals for the Federal Circuit, 99-1323, *In Re Independent Service Organizations Antitrust Litigation, CSU, et. al. v. Xerox Corporation,* Decided February 17, 2000, Judge Mayer writing the opinion.

19. For a discussion of self help focusing on copyright holders, see Dam 1998.

20. See the June 26, 1997 press release at http://www.usdoj/gov/atr/public/press_releases/1997/1173.htm.

21. See the December 17, 1998 press release at http://www.usdoj.gov/atr/public/press_releases/1998/2120.htm.

22. See the June 10, 1999 business review letter at http://www.usdoj.gov/atr/public/press_releases/1999/2484.htm.

23. See the March 24, 1998 press release at http://www.ftc.gov/opa/1998/9803/eye.htm.

24. For a description of the settlement, see the August 21, 1998 press release at http://www.ftc.gov/opa/1998/9808/sumvisx.htm. Despite this settlement, the FTC continued to pursue VisX for allegedly acquiring a key patent by inequitable conduct and fraud by omission on the U.S. Patent and Trademark Office. However, an administrative law judge subsequently dismissed this complaint; see the June 4, 1999 press release at http://www.ftc.gov/opa/1999/9906/visx.htm.

25. Note that these rules can create the perverse incentive for patent holders to assert that at least some of their patents are not in fact essential, but perhaps merely extremely helpful, in complying with the standard. By this device, a patent holder can in principle either refuse to license its patent to others (especially once the standard has become established, and perhaps for a patent that issued after the standard is established) or seek something more than fair and reasonable royalties. Of course, whether the terms fair and reasonable are evaluated on an ex ante or ex post basis is not precisely clear, although the terms would have little force if applied only on an ex post basis.

26. "Second Amended Complaint," *Disctronics Texas, Inc., et al. v. Pioneer Electronic Corp. et al.* Eastern District of Texas, Case No. 4:95 CV 229, filed August 2, 1996 at 12.

27. Note that a company might profit from *refusing* to participate in the standard setting process, in the hope that the resulting standard will nonetheless (perhaps inadvertently) infringe on the company's patent. Then the company would not be obligated to license its blocking patent on fair and reasonable terms, if at all. This would at least create the possibility that the company in question could control the standard and make it proprietary once it became established.

28. See Anton and Yao 1995 for a more complete discussion of the legal treatment of performance standards.

29. *Allied Tube & Conduit Corp. v. Indian Head, Inc.,* 486 U.S. 492, 500 (1988).

30. Federal Trade Commission. 1996, June. "Anticipating the 21st Century: Competition Policy in the New High-Tech Global Marketplace," Chapter 9, "Networks and Standards."

31. *ibid,* Chapter 10, "Joint Ventures," at 17.

32. There are many more examples of disputes involving hidden patent rights and standard setting, including: *Wang vs. Mitsubishi;* Microsoft and Cascading Style Sheets; and ETSI and Third-Generation Mobile Telephones.

33. See http://www.ftc.gov/opa/9606/dell2.htm.

34. I served as an expert in this matter retained by Rockwell; the views stated here do not necessarily reflect those of any party to the case.

35. See the May 3, 1995 press release at http://www.ftc.gov/opa/1995/9505/boscvis.htm. The recent merger of Gemstar and TV Guide is another example of a merger/settlement that raises antitrust issues.

36. One episode under investigation involves Abbott Laboratories, Novartis's Geneva Pharmaceuticals unit, and the popular hypertension drug, Hytrin. Another episode involves Aventis (the new company formed from the merger of Hoechst and Rhone-Poulenc), Andrx, and the heart drug Cardizem CD. Abbott reportedly agreed to pay Geneva $4.5 million per month to delay the launch of a generic version of Hytrin. Abbott asserts that its agreement with Geneva is "in accordance with all laws." See the *Wall Street Journal,* February 7, 2000, "FTC Panel Backs Suit Against Abbott, Novartis on Deal for Hypertension Drug," p. B20. See the FTC website for updates.

References

Anton, James, and Dennis Yao. 1995. "Standard-Setting Consortia, Antitrust, and High-Technology Industries." *Antitrust Law Journal* 64:247–65.

Baker, Jonathan B. 1999. "Promoting Innovation Competition Through the *Aspen/Kodak* Rule." *George Mason Law Review* 7:495–521.

Balto, David. 1999. "Networks and Exclusivity: Antitrust Analysis to Promote Network Competition." *George Mason Law Review* 7:523–76.

Cohen, Wesley M., Richard R. Nelson, and John Walsh. 2000, February. "Protecting Their Intellectual Assets: Appropriability Conditions and Why U.S. Manufacturing Firms Patent (or Not)." NBER Working Paper No. W7552.

Dam, Kenneth. 1998, August. "Self-Help in the Digital Jungle." John M. Olin Law & Economics Working Paper no. 59, University of Chicago Law School, Chicago.

Farrell, Joseph and Michael Katz. 1998. "The Effects of Antitrust and Intellectual Property Law on Compatibility and Innovation." *Antitrust Bulletin.*

Federal Trade Commission. 1996, May. "Competition Policy in the New High-Tech Global Marketplace." Washington DC. Staff Report.

Gilbert, Richard, and Carl Shapiro. 1997. "Antitrust Issues in the Licensing of Intellectual Property: The Nine No-No's Meet the Nineties." *Brookings Papers on Economics: Microeconomics:* 283–336.

Grindley, Peter, and David J. Teece. 1997. "Managing Intellectual Capital: Licensing and Cross-Licensing in Semiconductors and Electronics." *California Management Review* 39(2):1–34.

Hall, Bronwyn, and Rose Marie Ham. 1999. "The Patent Paradox Revisited: Determinants of Patenting in the U.S. Semiconductor Industry, 1980–94." NBER Working Paper No. W7062.

Heller, M. A., and R. S. Eisenberg. 1998. "Can Patents Deter Innovation? The Anticommons in Biomedical Research." *Science* 280:698–701.

Katz, Michael, and Carl Shapiro. 1985, Winter. "On the Licensing of Innovations." *Rand Journal of Economics.*

Katz, Michael, and Carl Shapiro. 1994. "Systems Competition and Network Effects." *Journal of Economic Perspectives* 8(2):93–115.

Klein, Joel I. 1997. "Cross-Licensing and Antitrust Law." Available at http://www.usdoj.gov/atr/public/speeches/1123.htm.

Kortum, S., and J. Lerner. 1998. "Stronger Protection or Technological Revolution: What is Behind the Recent Surge in Patenting?" Carnegie-Rochester Conference Series on Public Policy, Vol. 48 (June 1998): 247–304.

Lemley, Mark, and David McGowan. 1998. "Legal Implications of Network Economic Effects." *California Law Review* 86:481–611.

Merges, Robert P. 1999. "As Many as Six Impossible Patents Before Breakfast: Property Rights for Business Concepts and Patent System Reform." *Berkeley Technology Law Journal* 14(2):577–615.

Shapiro, Carl. 1989. "Theories of Oligopoly Behavior." In R. Schmalensee and R. Willig, eds., *Handbook of Industrial Organization.* New York: Elsevier Science Publishers: 330–414.

Shapiro, Carl. 1996a. "Antitrust in Network Industries." Available at http://www.usdoj/gov/atr/public/speeches/shapir.mar.

Shapiro, Carl. 1999. "Exclusivity in Network Industries." *George Mason Law Review* 7:673–84.

Shapiro, Carl, and Hal R. Varian. 1998. *Information Rules: A Strategic Guide to the Network Economy.* Cambridge, MA: Harvard Business School Press.

U.S. Department of Justice and Federal Trade Commission. 1995, April. *Antitrust Guidelines for the Licensing of Intellectual Property.* Washington DC.

U.S. Department of Justice and Federal Trade Commission. 2000, April. *Antitrust Guidelines for Collaborations Among Competitors.* Washington, DC.

Technical Appendix

Here I show that prices can be well above monopoly levels if multiple firms have critical patents, all of which read on a single product. More precisely, if N firms each control a patent that is essential for the production of a given product, and if these N firms independently set their licensing fees, the resulting markup on that product is N times the monopoly markup.

Suppose that N firms, $i = 1, \ldots, N$, each own a patent that is essential to the production of a given product. For simplicity, let us think of there being a competitive industry that produces this product, buying and assembling the necessary components from each of these N firms. For this purpose we can think of firm i either as setting a license fee for the use of its patent, or as setting a price at which it will sell its essential component to the competitive assembly industry; the theory is identical either way.

The cost to firm i per unit (for making and selling its component or for licensing its patent to assemblers) is denoted by c_i. The price of component i (or the license fee charged by firm i) is denoted by p_i. The price of the product itself is denoted by p. In addition to paying royalties (or buying components), the assembly firms incur an assembly cost per unit equal to α. Competition at the "assembly" level ensures that $p = \alpha + \sum_{i=1}^{N} p_i$.

Demand for the product in question is denoted by $D(p)$. The absolute value of the elasticity of demand is given by $\epsilon \equiv -D'(p)p/D(p)$. In general, ϵ will vary with p.

I assume that the N firms set their component prices, equivalently their license fees, independently and noncooperatively. In other words, I look for the Nash Equilibrium in the prices $p_1, \ldots, p_N$. The profits for firm i are given by

$$\pi_i = D(p)(p_i - c_i).$$

The first-order condition for firm i is given by

$$\frac{d\pi_i}{dp_i} = D(p) + D'(p)(p_i - c_i) = 0.$$

Adding up across all i gives

$$D(p)N + D'(p)\sum_{i=1}^{N}(p_i - c_i) = 0.$$

which can be rewritten as

$$\sum_{i=1}^{N}\frac{(p_i - c_i)}{p} = -\frac{D(p)}{pD'(p)}N.$$

Using the definition of the elasticity of demand, and the fact that $p = \alpha + \sum_{i=1}^{N} p_i$, we have

$$\frac{p - (\alpha + \sum_{i=1}^{N} c_i)}{p} = \frac{N}{\varepsilon}. \tag{1}$$

In other words, the percentage markup over cost for the product in question is equal to N times the inverse of the elasticity of demand. In contrast, the standard monopoly markup rule would be

$$\frac{p - (\alpha + \sum_{i=1}^{N} c_i)}{p} = \frac{1}{\varepsilon}. \tag{2}$$

The markup with N independent firms controlling key patents is equal to N times the monopoly markup.

It can be shown that the combined profits of the N firms under independent pricing is lower than would be earned by a monopolist selling all N components. This implies that the firms have an incentive to coordinate their pricing. A package license for all N components would lead to higher (combined) profits and lower prices for consumers.

5

Commercialization of the Internet: The Interaction of Public Policy and Private Choices or Why Introducing the Market Worked So Well

Shane Greenstein, *Kellogg Graduate School of Management, Northwestern University and NBER*

Executive Summary

Why did commercialization of the Internet go so well? This paper examines events in the Internet access market as a window on this broad question. The study emphasizes four themes. First, commercializing Internet access did not give rise to many of the anticipated technical and operational challenges. Entrepreneurs quickly learned that the Internet access business was commercially feasible. Second, Internet access was malleable as a technology and as an economic unit. Third, privatization fostered attempts to adapt the technology in new uses, new locations, new market settings, new applications and in conjunction with other lines of business. These went beyond what anyone would have forecast by examining the uses for the technology prior to 1992. Fourth, and not trivially, the NSF was lucky in one specific sense. The Internet access industry commercialized at a propitious moment, at the same time as the growth of an enormous new technological opportunity, the World Wide Web. As it turned out, the web thrived under market oriented, decentralized, and independent decision making. The paper draws lessons for policies governing the commercialization of other government managed technologies and for the Internet access market moving forward.

I. Motivation

The "commercialization of the Internet" is shorthand for three nearly simultaneous events: the removal of restrictions by the National Science Foundation (NSF) over use of the Internet for commercial purposes, the browser wars initiated by the founding of Netscape, and the rapid entry of tens of thousands of firms into commercial ventures using technologies which employ the suite of TCP/IP standards. These events culminated years of work at NSF to transfer the Internet into commercial hands from its exclusive use for research activity in government funded laboratories and universities.

Sufficient time has passed to begin to evaluate how the market performed after commercialization. Such an evaluation is worth doing. Actual events have surpassed the forecasts of the most optimistic managers at NSF. Was this due to mere good fortune or something systematic whose lessons illuminate the market today? Other government managed technologies usually face vexing technical and commercial challenges that prevent the technology from diffusing quickly, if at all. Can we draw lessons from this episode for the commercialization of other government managed technologies?

In that spirit, this paper examines the Internet access market and one set of actors, Internet Service Providers (ISPs). ISPs provide Internet access for most of the households and business users in the country (NTIA 1999), usually for a fee or, more recently, in exchange for advertising. Depending on the user facilities, whether it is a business or a personal residence, access can involve dial-up to a local number or 1-800 number at different speeds, or direct access to the user's server employing one of several high speed access technologies. The largest ISP in the United States today is America-On-Line, to which approximately half the households in the U.S. subscribe. There also are many national ISPs with recognizable names, such as AT&T Worldnet, MCI WorldCom/UUNet, Mindspring/Earthlink, and PSINet, as well as thousands of smaller regional ISPs.

The Internet access market is a good case to examine. Facilities for similar activity existed prior to commercialization, but there was reason to expect a problematic migration into commercial use. This activity appeared to possess idiosyncratic technical features and uneconomic operational procedures which made it unsuitable in other settings. The Internet's exclusive use by academics and researchers fostered cautious predictions that unanticipated problems would abound and commercial demand might not materialize.

In sharp contrast to cautious expectations, however, the ISP market displayed three extraordinary features. For one, this market grew *rapidly*, attracting thousands of entrants and many users, quickly achieving mass-market status. Second, firms offering this service became *nearly geographically pervasive*, a diffusion pattern rarely found in new infrastructure markets. And third, firms *did not settle* on a standard menu of services to offer, indicative of new commercial opportunities and also a lack of consensus about the optimal business model for this opportunity. Aside from defying expectations, all three traits—rapid growth, geographic pervasiveness, and the absence of settlement—do not inherently go together in most markets. The presence of restructuring should have interfered with rapid growth and geographic expansion. So explaining this market experience is also interesting in its own right.

What happened to make commercialization go so well? This paper's examination reveals four themes. First, commercialization did not give rise to many of the anticipated technical and operational challenges. Entrepreneurs quickly learned that the Internet access business was commercially feasible. This happened for a variety of economic reasons. ISPs began offering commercial service after making only incremental changes to familiar operating procedures borrowed from the academic setting. It was technically easy to collect revenue at what used to be the gateway functions of academic modem pools. Moreover, the academic model of Internet access migrated into commercial operation without any additional new equipment suppliers.

Second, Internet access was malleable as a technology and as an economic unit. This is because the foundation for Internet interconnectivity, TCP/IP, is not a single invention, diffusing across time and space without changing form. Instead, it is embedded in equipment that uses a suite of communication technologies, protocols, and standards for networking between computers. This technology obtains economic value in combination with complementary invention, investment, and equipment. While commercialization did give rise to restructuring of Internet access to suit commercial users, the restructuring did not stand in the way of diffusion, nor interfere with the initial growth of demand.

Third, privatizing Internet access fostered customizing Internet access technology to a wide variety of locations, circumstances, and users. As it turned out, the predominant business model was feasible at small scale and, thus, at low levels of demand. This meant that the technology was commercially viable at low densities of population, whether or not it was part of a national branded service or a local geographically concentrated service. Thus, privatization transferred the operation of the technology to a new set of decision makers who had new ideas about what could be done with it. Since experimentation was not costly, this enabled attempts to adapt the technology in new uses, new locations, new market settings, new applications, and in conjunction with other lines of business. While many of these attempts failed, a large number of them also succeeded. These successes went well beyond what anyone would have forecast by examining the limited uses for the technology by noncommercial users prior to 1992.

Fourth, and not trivially, the NSF was lucky in a particular sense of the word. It enabled the commercialization of the Internet access industry at a propitious moment, at the same time as the growth of an enormous new technological opportunity, the World Wide Web. This invention motivated further experimentation to take advantage of the new opportunity, that, as it turned out, thrived under market oriented and decentralized decision making.

The paper first develops these themes. Then it describes recent experience. It ends by discussing how these themes continue to resonate today.

II.   Challenges During Technology Transfer: An Overview

Conventional approaches to technological development led most observers in 1992 to be cautious about the commercialization of the Internet. To understand how this prediction went awry, it is important to understand its foundations.

Many studies of the commercialization of technology emphasize the situated nature of technological development. Technologies do not simply spring out of the ether; instead, learning processes and adaptation behavior shape them. Users and suppliers routinely tailor technologies to short term needs, making decisions that reflect temporary price schedules or idiosyncratic preferences, resulting in technological outcomes that can only be understood in terms of these unique circumstances and origins.[1] Such themes resonate throughout studies of technologies which develop under government management.[2]

Seen through this light, the most problematic feature of the Internet was its long exclusive use by military, government, or academic users. Prior to 1992 it had developed into the operations found at an academic modem pool or research center. These were small scale operations, typically serving no more than several hundred users, involving a mix of frontier and routine hardware and software. A small operation required a server to monitor traffic and act as a gatekeeper, a router to direct traffic between the Internet and users at PCs within a local-area-network (LAN) or calling center, and a connection to the Internet backbone or data exchange point operated by the NSF. These were often run by a small staff, either students or information technology professionals.

Revenues were not regularly collected in these arrangements and budgetary constraints were not representative of what might arise with commercial operations and competitive pressures. Many small colleges had opened their Internet connections with NSF subsidies. The organizational arrangement within research computing centers also was idiosyncratic, usually with only loose ties, if any, to the professionally run administrative computing centers of a university or research organization. The array of services matched the needs of academic or research computing, which had only a partial overlap with the needs of commercial users.

Any student of technology transfer would have confidently predicted that the transition into commercial markets would give rise to challenges. Standing in 1992 and looking forward, it was uncertain whether these challenges would take a long time to solve and whether commercial users' needs would be difficult to address. In general, conventional analysis anticipates one of three challenges: *technical, commercial*, and *structural challenges.*

Technical challenges often arise during commercialization. Government users, government procurement, and government subsidies result in technology with many features mismatched to commercial needs. Products possessed features for which vendors or users have no need. Alternatively, commercial vendors and users do need other features. Thus, as a technical or engineering matter, a technology which is mature for exclusive noncommercial uses—such as a military application—may appear primitive in civilian use. It may require complementary inventions to become commercially viable. If these requirements are considerable, then commercialization may occur slowly.

For example, military users frequently require electronic components to meet specifications that suit the component to battle conditions. Extensive technical progress is needed to tailor a product design to meet these requirements. Yet, and this is difficult to anticipate prior to commercialization, an *additional* amount of invention is often needed to bring its manufacturing to a price/point with features that meet more cost-conscious or less technically stringent commercial requirements.

Commercial challenges arise when commercial markets require substantial adaptation of operation and business processes in order to put technologies into use. In other words, government users or users in a research environment often tolerate operational processes that do not translate profitably to commercial environments. After a technology transfers out of government sponsorship, it may not be clear how to balance costs and revenues for technologies that had developed under settings with substantial subsidies underwriting losses, and research goals justifying expenditures. Hence, many government managed technologies require considerable experimentation with business models before they begin to grow, if they grow at all.

For example, the supersonic transport actually met its engineering targets, but still failed to satisfy basic operational economics in most settings. Being technically sleek was insufficient to attract enough interest to generate the revenue to cover operating costs on any but a small set of routes. No amount of operational innovations and marketing campaigns were able to overcome these commercial problems.

New technologies are also vulnerable to *structural challenges* that impede pathways to commercialization. Commercial and structural challenges are not necessarily distinct, though the latter are typically more complex. Structural challenges are those that require change to the bundle of services offered, change to the boundary of the firms offering or using the new technology, or dramatic change to the operational structure of the service organization. These challenges arise because technologies developed under government auspices may presume implementation at a particular scale or with a set of technical standards, but require a different set of organizational arrangements to support commercial applications.

For example, while many organizations provided the technical advances necessary for scientific computing in academic settings during the 1950s, very few of these same firms migrated into supporting large customer bases among business users. As it turned out, the required changes were too dramatic for many companies to make. The structure of the support and sales organization were very different, and so too were the product designs. Of course, the few who successfully made the transition to commercial users, such as IBM, did quite well, but doing so required overcoming considerable obstacles.

In summary, conventional analysis forecasts that migrating Internet access into commercial use would engender technical, commercial, and structural challenges. Why did the migration proceed so different from what was expected?

III.   The Absence of Challenge in the Internet Access Industry

An ISP is a commercial firm that provides access, maintains it for a fee, and develops related applications as users require. While sometimes this is all they do, with business users they often do much more. Sometimes ISPs do simple things such as filtering. Sometimes it involves managing and designing e-mail accounts, databases, and web pages. Some ISPs label this activity consulting and charge for it separately; others do not consider it distinct from the normal operation of the Internet access services.

On the surface the record of achievement for ISPs is quite remarkable. Most recent surveys show that no more than 10% of U.S. households get their Internet access from university sponsored Internet access providers, the predominant provider of such access prior to commercialization. Today almost all users go to a commercial provider (Clemente 1998, Nie and Ebring 2000). As of 1997, this ISP industry was somewhere between a three and five billion dollar industry (Maloff 1997), and it is projected to be much larger in a few years.

By the end of the century the ISP market had obtained a remarkable structure. One firm, America On-Line, provided access to close to half the households in the U.S. market, while several score of other ISPs provided access to millions of households and businesses on a nationwide basis. Thousands of ISPs also provided access for limited geographic areas, such as one city or region. Such small ISPs accounted for roughly a quarter of household use and another fraction of business use.

Technical Challenges Did Not Get in the Way

The Internet access market did suffer from some technical challenges, but not enough to prevent rapid diffusion. Commercialization induced considerable technical innovation in complementary inventive activities. Much of this innovative activity became associated with developing new applications for existing users and new users.

It is often forgotten that when the electronic commerce first developed based on TCP/IP standards, it was relatively mature in some applications, such as e-mail and file transfers, which were the most popular applications (these programs continue to be the most popular today, NTIA 1999). To be sure, TCP/IP based programs were weak in other areas, such as commercial database and software applications for business use, but those uses did not necessarily have to come immediately. The invention of the World Wide Web in the early 1990s further stretched the possibilities for potential applications and highlighted these weaknesses.

More important for the initial diffusion, little technical invention was required for commercial vendors to put this technology into initial mainstream use. Academic modem pools and computing centers tended to use technologies similar to their civilian counterparts—such as bulletin board operators—while buying most equipment from commercial suppliers. Moving this activity into the mainstream commercial sector did not necessitate building a whole new Internet equipment industry; it was already there, supplying goods and services to the universities and to home PC users. Similarly, much of the software continued to be useful—that is, Unix systems, the gatekeeping software, and the basic communication protocols. Indeed, every version of Unix software had been TPC/IP compatible for many years due to Department of Defense requirements. A simple commercial operation only needed to add a billing component to the gatekeeping software to turn an academic modem pool into a rudimentary commercial operation.

Technical information about these operations was easy to obtain if one had sufficient technical background; a BA in basic electrical engineering or computer science was far more than adequate. Many ISP entrepreneurs had used the technology as students or in related lines of business. Descriptions of some of the earliest access operations show that they did not employ any exotic hardware or rare technologies (Kalakota and Whinston 1996, Kolstad 1998). Many Internet bulletin boards quickly developed and Boardwatch Magazine, among others, expanded its focus from bulletin boards to ISP as early as 1994, also spreading information about how to operate such ventures. Several vendor associations, such as the Commercial Internet Exchange, were formed and also served as information sources.

Users with investments in networking technology, such as LANs or simple client/server architectures, also could adopt basic features with little further invention. Internet technologies associated with textual information had incubated for 20 years and were well past the necessary degree of technical maturity necessary for mainstream use. Telnet, FTP, and the basic protocols for e-mail were widely diffused and relatively easy to use. Some communication software already used TCP/IP and many of the common programs could easily adapt to it. There were already many similar technical activities taking place in commercial settings. TCP/IP compatibility was built into Windows 95, which further eased investments for users after 1995.

The basic commercial transaction for Internet access also did not raise prohibitive technical issues. Most often it involved repetitious and ongoing transactions between vendor and user. A singular transaction arose when the vendor performed one activity, setting up Internet access or attaching Internet access to an existing computing network. If the ISP also operated the access for the user, then this ongoing operation provided frequent contact between the user and vendor, and it provided frequent opportunity for the vendor to change the delivery of services in response to changes in technology and changes in user needs. This worked well because in many cases an ISP was better educated about the technological capabilities than the user. In effect, the ISP sold that general knowledge to the user in some form that customized it to the particular needs and requirements of the user. At its simplest level, this provided users with their first exposure to a new technological possibility while educating them about its potential.

Often access went beyond exposure to the Internet, especially with a business user, and included the installation, maintenance, and training, as well as application development. These types of transfers of knowledge typically involved a great deal of nuance, often escaped attention, and yet were essential to developing infrastructure markets as an ongoing and valuable economic activity. The basic technical know-how did not differ greatly from routine knowledge found in the computing services sector prior to commercialization.

Finally, some NSF decisions and legacy regulatory decisions also aided. When the NSF took over stewardship of the Internet backbone, it invested in developing a scalable system of address tables and IP-address systems. Subsequent growth tested those investments and inventions; no surprising problems were found, nor did any engineering problems hinder growth. Domain name registration also remained a gentle monopoly until recently. Data exchange points remained organized around the cooperative engineering principles used within the NSF days. A competitive data communications industry was beginning to reach adolescence at about the same time as commercialization and provided additional access points for new firms, particularly in urban areas. So as a technical matter, interconnection with the public switch network did not pose any significant engineering challenges (Werbach 1997).

Commercial Challenges Did Not Slow Diffusion

Internet access was built in an extremely decentralized market environment. Aside from the loosely coordinated use of a few de facto standards (such as the World Wide Web consortium) government mandates after commercialization were fairly minimal. ISPs had little guidance or restrictions. They were therefore able to tailor their offerings to local market conditions and to follow entrepreneurial hunches about growing demand.

As a technical matter, there were few barriers to entry in the provision of dial-up access. As a result, commercial factors, and not the distribution of technical knowledge among providers, largely determined the patterns of development of the basic dial-up access market immediately after commercialization. To the surprise of many, the operational procedures developed over two decades lent themselves to the early commercial implementations, fostering a foundation for commercial growth. As with many new markets which spawn in noncommercial environments (Ventresca et al. 1998), many features were borrowed wholesale and without question. In effect, entrepreneurs borrowed the organization of the academic modem pool and tried to put a revenue generating function on top of it. Billing software was added to the basic gateway component, and once this proved to be a feasible way to collect revenue, many entrepreneurs built on top of that commercial form.

Shortly after commercialization in 1994, only a few commercial enterprises offered national dial-up networks with Internet access, mostly targeting the major urban areas. Pricing was not standardized and varied widely (Boardwatch 1994–1995). Most of these ISPs were devoted to recreating the type of network found in academic settings or modifying a commercial bulletin board with the addition of backbone connections, so interconnection among these firms did not raise insoluble contracting or governance problems. These ISPs were devoted primarily to dial-up; few ISPs attempted sophisticated data transport over higher speed lines, where the regulatory issues could be more complex and where local exchange competitors were developing the nascent market.

Very quickly ISPs learned that low cost delivery required locating access facilities close to customers. This had to do with telephony pricing policies across the U.S. The U.S. telephone system has one pervasive feature; distance-sensitive pricing at the local level. In virtually every part of the country, phone calls over significant distances (i.e., more than 30 miles) engender per minute expenses, but local calls are usually free. Hence, Internet access providers had a strong interest in reducing expenses to users by providing local coverage. Unmet local demand was a commercial opportunity for an entrepreneurial ISP.

As it turned out, access over dial-up lent itself to small scale commercial implementations. Several hundred customers could generate enough revenue to support physical facilities and a high-speed backbone connection in one location, so scale economies were not very binding. The marginal costs of providing dial-up services were low and the marginal costs of expansion also fell quickly, as remote monitoring technology made it cheap to open remote facilities. The marginal costs to users of dial-up service were also low in response, involving only incremental changes for organizations that had experience with PC use or LAN technology. It was easy to generate revenue in subscription models, where a commercial firm withheld availability of access unless payment was made. Hence, the economic thresholds for commercial dial-up service turned out to be feasible on a very small scale, encouraging small firms and independent ISPs. To be sure, many firms also tried to implement access businesses on a large scale, but the economic advantage of large scale did not preclude the entry of small scale firms, at least not at first.

Finally, decades of debate in telephony had already clarified many regulatory rules for interconnection with the public switch network, eliminating some potential local delays in implementing this technology on a small scale. The FCC treated ISPs as an enhanced service, not passing on access charges to them as if they were competitive telephone companies, effectively making it cheaper and administratively easier to be an ISP. This decision did not receive much notice at the time since most insiders did not anticipate the extent of the growth that would arise. As ISPs have grown and as they threaten to become competitive voice carriers, these interconnection regulations have come under more scrutiny (Sidek and Spulber 1998, Weinberg 1999).[3]

In retrospect, two key events of 1995 set the stage for the commercial ISP market for the remainder of the decade. The first was the Netscape IPO in August 1995. The other was the entry of AT&T World Net.

The World Wide Web was known in the academic community in the early 1990s. It began to diffuse prior to commercialization and accelerated with Mosaic, a prototype browser developed at the University of Illinois. Many ISPs included Mosaic on their systems. Despite licensing the technology to many firms, the University of Illinois did not generate as much excitement as the Netscape IPO, which brought extensive publicity to the new technology (Cusumano and Yoffie 1998). The subsequent browser wars further heightened this awareness.

The emergence of the web changed the commercial opportunities for ISPs. ISPs found themselves both providing a traditional service in demand, text-based applications such as e-mail, and trying to position themselves for a new service, web applications. This new opportunity provided strong incentives to grow and experiment with new business models and new lines of service. It also induced considerable new entry. While not all markets experienced the same type of competitive choices, nor did all ISPs see the same opportunities, many private firms found ways to develop opportunities quickly, learning lessons that they then applied in other localities.

AT&T's entry was also important but its actions mattered because of what did not happen rather than what did. AT&T developed a nationwide Internet access service, which was available in much of the country, opening with as large a geographic spread as any other contemporary national provider. It also grew quickly, acquiring one million customers with heavy publicity and marketing. This growth depended on the strength of its promise to be reliable, competitively priced, and easy to use. It was deliberately aimed at households, and provided a mass-market service from a name brand. It was a commercial success, to be sure, but that was all. It was not a huge or dominant success, nor did it initiate a shakeout or restructuring of the market for ISP service.

Here was a branded, nationwide, professionally operated subscription model of ISP service, opening with as large a geographic spread as any other contemporary national provider. Yet, it did not end the growth of others, such as AOL, nor did it stop new entry of small firms, such as Mindspring, nor did it initiate a trend toward consolidation around a few national branded ISP services. In other words, even with its deep pockets AT&T did not dominate the offerings from all other firms, nor did it end the restructuring of the access business. This defied many predictions about how this market would be structured, further encouraging the decentralized growth and the emergence of independent ISPs.

Growth and entry brought about extraordinary results. Downes and Greenstein (1998) have constructed maps that illustrate the density of location of ISPs at the county level for the fall of 1996 and 1998; black and white versions of these are shown in figures 5.1 and 5.2.[4] For color versions see, respectively: http://www.kellogg.nwu.edu/faculty/greenstein/images/htm/Research/Maps/mapsep1.pdf and http://www.kellogg.nwu.edu/faculty/greenstein/images/htm/Research/Maps/mapoct98.pdf.

Colored areas are countries with providers. White areas have none. As the maps show, ISPs tend to locate in all the major population cen-

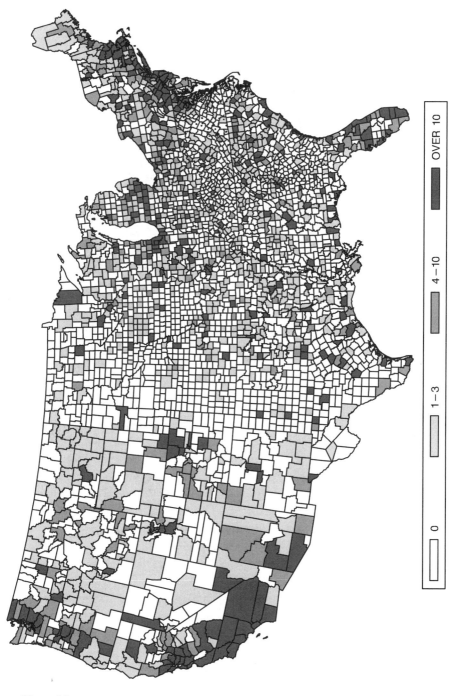

Figure 5.1
Distribution of ISPs September 1996
Copyright © 1998 Tom Downes and Shane Greenstein.

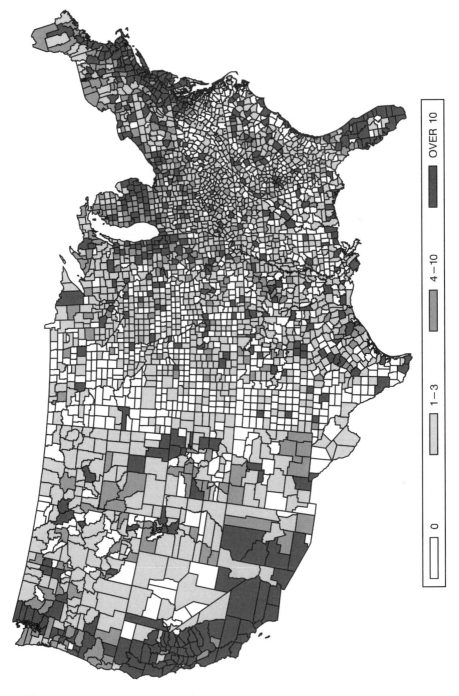

Figure 5.2
Distribution of ISPs October 1998
Copyright © 1998 Tom Downes and Shane Greenstein.

ters, but there are also plenty of providers in rural areas. The maps also illustrate the importance of changes over time. Many of the areas that had no coverage in the fall of 1996 were covered by the fall of 1998. Many of the areas that had competitive access markets in the early period were extraordinarily competitive in the latter period.

Downes and Greenstein (1998) show that more than 92% of the U.S. population had access by a short local phone call to seven or more ISPs by 1998. No more than 5% did not have any access. Almost certainly the true percentage of the population without access to a competitive dial-up market is much lower than 5%. In other words, with the notable exception of some low-density areas, ISP service was quickly available everywhere. To put it simply, among the vast majority of the U.S. population in urban and suburban areas, lack of use was primarily due to demand factors, not the absence of supply.

An unexpected pattern accompanied this rapid growth in geographic coverage. First, the number of firms maintaining national and regional networks increased over the 2 years. In 1996, most of the national firms were recognizable; they were such firms as IBM, AT&T, and other established firms who entered the ISP business as a secondary part of their existing services, such as providing data services to large corporate clients. AOL, CompuServe, and Prodigy all were in the process of converting their online service, previously run more like bulletin boards than ISPs, into Internet providers. By 1998, many entrepreneurial firms maintained national networks and few of these new firms were recognizable to anyone other than an industry expert.

There was also a clear dichotomy for growth paths of entrepreneurial firms who became national and regional firms. National firms grow geographically by starting with major cities across the country and then progressively moving to cities of smaller populations. Firms with a regional focus grow into geographically contiguous areas, seemingly irrespective of urban or rural features.[5]

Most of the coverage in rural areas comes from local firms. In 1996, the providers in rural counties with under 50,000 population were overwhelmingly local or regional. Only for populations of 50,000 or above do national firms begin to appear. In the fall of 1998, the equivalent figures were 30,000 or lower, indicating that some national firms had moved into slightly smaller areas and less dense geographic locations. In other words, Internet access in small rural towns is largely done by local or regional providers, with national firms only slowly expanding into similar territory.

It appears as if it does not pay for many large national providers to provide dial-up service for the rural areas whereas many small local firms in other lines of business (e.g., local PC retailing) can afford to add Internet access to their existing business. It may also be the case that the local firm may have an easier time customizing the Internet access business to the unique needs of a set of users in a rural setting.

What Structural Challenges Arose?

Commercialization of the Internet created an economic and business opportunity for providing access. The costs of entry into low quality dial-up access were low, and commercially oriented firms filled voids in specific places. For any firm with national ambitions, coverage of the top 50 to 100 cities in the U.S. was a fleeting advantage and quickly become a necessity for doing business. For any local or regional firm in an urban market, many competitors arose.

Yet, not long after the Netscape IPO the ISP industry began to enter a second phase. Profitability and survival involved more than geographic expansion. It involved bringing ISP service to the households and businesses with PCs, but without access. It also involved expanding into services which took advantage of new opportunities associated with the web.

Understanding this second phase requires an understanding of the services ISPs offer other than basic access and how those began to evolve. These new services include one of several activities: monitoring technical developments, distilling new information into components that are meaningful to unfamiliar users, and matching unique user needs to one of many new possible solutions enabled by advancing technical frontiers. Sometimes it includes heavy use of the technological frontier and sometimes not. In general, it depends on the users, their circumstances, their background, their capital investments, the costs of adjusting to new services, and other factors that influence the match between user needs and technological possibilities.

ISPs commercialized their adaptive role by offering new services that can be grouped into five broad categories: networking, hosting, web page design, basic access, and frontier access (see the appendix of Greenstein 1999 for precise definitions).

Networking involves activities associated with enabling Internet technology at a user's location. All ISPs do a minimal amount of this as part of their basic service in establishing connectivity. However, an extensive array of these services, such as regular maintenance, assessment of facilities, emergency repair, and so on, are often essential to keeping and retaining business customers. Note, as well, that some of these services could have been in existence prior to the diffusion of Internet access.

Hosting is typically geared toward a business customer, especially those establishing virtual retailing sites. This requires the ISP to store and maintain information for its access customers on the ISP's servers. All ISPs do a minimal amount of hosting as part of basic service, even for residential customers (e.g., for e-mail). However, some ISPs differentiate themselves by providing an extensive array of hosting services, including credit card processing, site analysis tools, and so on.

Web design may be geared toward either the home or business user. Again, many ISPs offer some passive assistance or help pages on web page design and access. However, some offer additional extensive consulting services, design custom sites for their users, and provide services associated with design tools and web development programs. Most charge fees for the additional services.

Basic access constitutes any service as slow as or slower than a T-1 line. Many of the technologies inherited from the precommercial days became standard parts of basic access and were not regarded as a new service. A number of other new functions, such as audio streaming, filtering, and linking, also gradually became standard parts of most firms' offerings. Frontier access includes any access faster than a T-1 line, which is becoming the norm for business access. It also includes ISPs that offer direct access for resale to other ISPs or data carriers and ISPs that offer parts of their own backbone for resale to others.[6]

By 1998, different ISPs had chosen different approaches, offering distinct combinations of services and distinct geographic scopes. Table 5.1 shows the results of a survey of the business lines of 3,816 Internet service providers in the United States who advertise on *thelist*, an online directory of ISPs, in the summer of 1998 (see the appendix of Greenstein 1999). Virtually every firm in the sample provides some amount of dial-up or direct access and basic functionality, such as e-mail accounts, shell accounts, IP addresses, new links, FTP, and Telnet capabilities, but these 3,816 seem to underrepresent both very small and quasi public ISPs (e.g., rural telephone companies).

Table 5.1
Product Lines of ISPs

Category Definition	Most Common Phrases in Category	Original Sample
Providing and servicing access though different channels	28.8, 56 k, ISDN, web TV, wireless access, T-1, T-3, DSL, frame relay, e-mail, domain registration, new groups, real audio, FTP, quake server, IRC, chat, video conferencing, cybersitter TM	3,816 (100%)
Networking service and maintenance	Networking, intranet development, WAN, colocation server, network design, LAN equipment, network support, network service, disaster recovery, backup, database services, Novell Netware, SQL server	789 (20.6%)
Web site hosting	Web hosting, secure hosting, commercial site hosting, virtual FTP server, personal web space, web statistics, BBS access, catalog hosting	792 (20.7%)
Web page development and servicing	Web consulting, active server, web design, Java, perl, VRML, front page, secure server, firewalls, web business solutions, cybercash, shopping cart, Internet marketing, online marketing, electronic billing, database integration	1,385 (36.3%)
High speed access	T-3, DSL, xDSL, OC3, OC12, Access rate > 1056 k	1,059 (27.8%)

Of the 3,816 ISPs, 2,295 (60.1%) have at least one line of business other than basic dial-up or direct Internet access. Table 5.1 shows that 1,059 provide high speed access, 789 networking, 792 web hosting, and 1,385 web page design. There is some overlap: 1,869 do at least one of either networking, hosting, or web design; 984 do only one of these three; 105 do all three as well as frontier access. This reveals many different ways to combine nonaccess services with the access business.[7]

The Contours of Response to Structural Challenges

Structural issues were not resolved quickly and have not disappeared as of this writing. This occurred because these activities contain much more complexity and nuance than table 5.1 can display.

ISPs customize Internet technologies to the unique needs of users and their organizations, solving problems as they arise, and tailoring general solutions to idiosyncratic circumstances and their particular commercial strengths. Sometimes ISPs call this activity consulting, and charge for it separately; sometimes it is included as a normal business practice. In either case, it involves the translation of general knowledge about Internet technologies into specific applications that yield economic benefits to end users.

What factors influenced vendors' attempts to construct viable and ongoing economic entities using new technology in an evolving market place? Is it possible to classify and analyze the determinants of coinvention? Why did some regions play host to ISP growth and others did not? There are many explanations, but these aggregate into two classes, one which emphasizes firm specific factors and another which emphasizes location specific factors.

Firm Specific Factors   Firm specific factors shape the incentives to bring new technology into use (see, e.g., Demsetz 1988 or Nelson and Winter 1977 for a summary). ISPs came to the new opportunities with different skills, experiences, or commercial focus. In the face of considerable firm specific commercial uncertainty, ISPs purchased and installed their own capital equipment, publicized brand and service agreements, and made other long-lasting investments. Many of these investments could commit the ISP to particular services, even before market demand was realized or new commercial opportunities were recognized.

Strategies pursued by national firms can be viewed in this light. Most national ISPs covered the same geographic territories, so their strategies reflected either unique assets at the firm level, a firm's vision for where their service should fall relative to competitors, or some other firm specific feature. A more detailed look at each of IBM, AT&T, AOL, Earthlink/Mindspring, and PSINet will illustrate the variety of strategies each pursued.

IBM had been an early entrant into the ISP market, focusing primarily on business customers and secondarily on home users. Their service grew rapidly nationwide and globally, complementing their considerable other computer services. Yet, in a few years the firm decided to divest itself of its ISP backbone and facilities, eventually selling to the highest bidder, AT&T. The firm concluded that joint provision of access and other computer services was not a strategic advantage, and therefore focused its attention on computer operations in many firms. The full benefits from this refocusing will only be manifest in time.

AT&T entered into consideration in another way. As already noted, it added a dial-up service soon after commercialization. In 1998 it purchased TCI/@home, a cable company, and Excite, a web portal. These acquisitions position them for providing data service to the home with some content. With the recent agreement to purchase Media One, which was pending at the FCC as of this writing, AT&T became the largest cable provider in the country. The benefits from this are somewhat speculative, as the revenue stream justifying these purchases has not been realized. If voice telephony, streaming media, or any other of the host of new broadband services become viable over cable lines, AT&T is well positioned to provide them. Subscription fees for high speed access could also justify these purchases, if that technology becomes widely adopted.

AOL took a different approach. First, it grew its home user base through aggressive marketing to less technical users. In response to the proliferation of ISPs in the mid 1990s, it ended its tiered subscription model and introduced a flat-rate pricing model which mimics these other ISPs. Next it bought CompuServe, a failed competitor with a loyal customer base, and currently operates it as a separate branded entity. It also sold off its access facilities to UUNet, a subdivision of MCI/Worldcom, announcing concentration on the development of content. It has since pursued its walled garden strategy of making AOL proprietary content attractive and the primary focus of AOL users. The purchase of ICQ, an instant messaging service, and Time/Warner, among others, are consistent with this strategic approach. It is still an ISP, but a unique one, providing access to the Internet that its customer base infrequently uses. The full benefits of this approach are speculative as of this writing, as the revenues from it have not been fully realized.

Earthlink and Mindspring illustrate the issues facing new entrants on a national level. They market a low-cost reliable service which is also easy to use, successfully competing against AT&T with much the same appeal but a different branding. These firms also specialize in making the Internet easy to use for the nonAOL user, the web surfer who wants some but not too much help. Eventually these firms merged, partly to consolidate their resources for competition against AOL, and partly to compete more strongly in the nonAOL customer space. As one of the largest dial-up services in the country, there is a big question whether they can survive in their niche in the face of competitive substitutes from all sides.

Finally, PSINet illustrates the feasibility of embarking on a strategy of emphasizing infrastructure. They started as a consumer Internet service, but got out of that business in 1996. They had built out their own backbone, investing in high speed facilities across the country, focused on becoming a carrier's carrier for other ISPs and for businesses. Part of their strategy involves heavy investments in complementary services, such as hosting services or corporate software services, that can offer high speed service when located next to fast Internet backbone lines. They also focus on offering infrastructure services to businesses, and developing services such as VPNs, which take advantage of their technical capabilities and nationwide coverage. Once again, the full benefits of this approach are speculative, depending on realizing demand in the future.

There are, of course, many other national firms. As with the above examples, their strategies mix different elements of speculative investment, restructuring of organizations, and entrepreneurial guesses about future demand. In all cases, these experiments involve executives making investments under technical and commercial uncertainty, restructuring production and distribution assets on a grand scale, trying to bring new services to market, and only finding out if they meet market demand years after those investments.

It is also important to recognize the variance associated with local and regional ISPs, another and particularly interesting subset of ISPs, that provide service for approximately between a fifth and a quarter of the Internet users in the U.S. These firms locate in many different parts of the country; hence their firm specific strategies are also influenced by factors associated with their locations.

Location Specific Factors   A well-known line of economic research, dating at least to Griliches (1957), has emphasized the geographic dispersion of incentives to adopt new technology. In this instance, while basic dial-up access is widely available in all urban areas and many rural areas (Downes and Greenstein 1998), there is great variance in market structure on a local level. Some areas contain many suppliers from a wide variety of backgrounds, while others contain few suppliers. From the standpoint of an ISP, many of these structural features of markets are exogenous, and shape the competitive pressures of the ISP. In addition, ISPs customize frontier technology to the needs of enterprises doing business at a specific time in a specific place. The costs of this may vary by region because infrastructure differs by region. The demand for higher speed service should also differ across regions if the users who find speed valuable are unevenly distributed across geographic regions—e.g., someone from San Francisco may be more willing to pay for speed than people from Poughkeepsie.

The contrast between firm specific and location specific questions are examined in Augereau and Greenstein (1999), who looked at small ISPs' investments in upgrades, and Greenstein (1999), who examined small ISPs' propensity to offer services beyond routine service associated with basic access. Both studies identify the importance of geographic factors by taking advantage of the variation between the locations of small ISPs.

These studies are motivated by two observations. First, as noted in Downes and Greenstein 1998, most large firms are located in the same (or largely overlapping) set of major cities. Hence, for the importance of location to be understood, the cause of variation between the small firms needs to be identified. Second, Greenstein (1999) and Strover (1999) document that ISPs in rural locations tend to provide fewer high quality services than those found in urban locations. Was this due to differences in infrastructure between urban and rural areas, differences in the type of customer found there, or differences in the types of entrepreneurs who locate in different regions?

These studies found that firm size, capacity, and financial strength were important determinants of behavior. There was also some evidence in Augereau and Greenstein 1999 that local infrastructure quality influenced investment behavior. Generally, variation in local demographic conditions or competitive conditions did not influence behavior. Both studies find much unmeasured variance in behavior, consistent with the presence of unmeasured location specific or firm specific determinants. Moreover, the factors which lead ISPs to offer new services, such as size, previous investments, and strategic focus, are disproportionately found in national firms and in local firms in urban areas.

These findings are consistent with the view that the scale of investment, the local infrastructure's quality, and the explicit costs shape investment decisions by young ISPs in emerging markets. It is also consistent with the view that there is too much commercial uncertainty in this market for firms to tailor the technical vintages of their capital stocks too closely to geographically local demand or competitive conditions. Finally, it is consistent with the view that most young firms with ambitious expansion plans initially locate in urban areas instead of rural areas, growing their base markets and expanding outward, if at all.

IV. Past Lessons and Future Challenges

As public discussion of electronic commerce has grown, a loose coalition of prophets for the new economy has come to dominate popular discussion. They write for such publications as *The Industry Standard, Business 2.0, Wired, Red Herring, Fast Company,* and more Webzines than anyone can list. It is only a slight exaggeration to say that all popular portrayals of the Internet contain two principal features. First, the prophets declare a business revolution in all information intensive activities—such as broadcasting, entertainment, retail marketing, supply chain management, other coordinative activity, and research. Second, and this is related, these same prophets proclaim that this technology's novelty dilutes standard lessons from the past. In other words, because this technology contains so many unique features, it is ushering in a new commercial era that operates according to new rules.

To be sure, there is probably a grain of truth to these declarations. However, momentary euphoria does not, nor should it, justify too simplistic a retrospective view of what actually happened, nor what is about to happen. Indeed, this paper showed that a traditional economic perspective does provide considerable insight into this new industry. In that spirit, we return to the questions that motivated the study and recap the findings.

The Commercialization of Internet Access Technology

Why Did the Internet Access Business Grow Quickly?   Stated simply, exclusive use did not lead to isolated technical and operational developments. Hence, commercializing Internet access did not give rise to any difficult or insolvable technical and operational challenges. This was due in no small part to the way in which the defense department and the NSF incubated the technology. It grew among researchers and academics without being isolated from commercial suppliers. That is, the technology grew without generating a set of suppliers whose sole business activity involved the supply of uniquely designed goods for military or government users. Related to this was the fact that the basic needs of researchers and academics were not so different from early commercial users. Hence, simple applications of the Internet invented for academic users—such as e-mail and file transfer using phone lines—migrated to commercial uses without much technical modification.

Why Did Geographic Ubiquity Arise?   To summarize, the Internet access business was commercially feasible at a small scale and, thus, at low levels of demand. This meant that the technology was commercially viable at low densities of population, whether or not it was part of a national branded service or a local geographically concentrated service. Again, this partly mimicked the academic experience, where the operations were also feasible on a small scale, but that statement alone does not capture all the factors at work. Internet access was feasible in a wide variety of organizational forms, large and small. Small scale business opportunities thrive with the help of entrepreneurial initiative that tends to be widespread throughout the U.S.—including many low density and isolated cities in otherwise rural areas that were largely not being served by national firms. Small scale implementation also depended on the presence of high quality complementary local infrastructure, such as digital telephony, and interconnection to existing communications infrastructure. These too were available throughout most of the U.S. due to national and local initiatives to keep the communications infrastructure modern.

Why Did the Internet Access Business Not Settle into a Common Pattern? Market forces did not impose uniformity in the use nor in the supply of access technology. Part of this was due to the absence of technical and commercial challenges, which allowed low cost experimentation of the technology in new uses, new locations, new market settings, new applications, and in conjunction with other lines of business. More generally, the technology was quite malleable as an economic unit. It could stand alone or become part of a wider and integrated set of functions under one organizational umbrella. Such malleability motivated experiments with new organizational forms for the delivery of access services, experiments which continue today. Finally, and unique to this example, the invention of the World Wide Web brought new promise to the technology. Not only did new business models arise to explore and develop its primitive capabilities and expand them into new uses, but it motivated firms to experiment with Internet access alongside new business lines.

Why Did Market Forces Lead to Such Extensive Growth?   This case illustrates how market forces can customize new technologies to users and implement new ways of delivering technologies. These activities have immense social value when there is uncertainty about technical opportunities and complex issues associated with implementation. In addition, as the literature on general purpose technology would put it, coinvention problems are best situated with those who face them. In this case, those actors were ISPs who knew about the unique features of the user, the location, or the application. More generally, commercialization transferred development into an arena where decentralized and unregulated decision making took over. This was precisely what was needed to customize Internet access technology to a wide variety of locations, circumstances, and users. Removing the Internet from the exclusive domain of NSF administrators and employees at research computing centers brought in a large number of potential users and suppliers, all pursuing their own vision and applying it to unique circumstances. In addition, it allowed private firms to try new business models, employing primitive web technologies in ways that nobody at the NSF could have imagined.

In What Sense Did the NSF Get Lucky?   As it turned out, the NSF commercialized the Internet access industry at a propitious moment, during the growth of an enormous new technological opportunity, the World Wide Web. Competitive forces sorted through new uses of this opportunity in particular places, enabling some businesses to grow and unsentimentally allowing unsuccessful implementations to fade. To be sure, some of these developments were heavily shaped by nonprofit institutions, such as the World Wide Web Consortium or the Engineering Internet Task Force, but profit motives still played a prominent role. Said another way, had NSF stewardship over the Internet continued there would have been some experimentation at computing centers found at universities and government laboratories, but it would not have been possible to replicate all the exploratory activity that did arise in commercial markets.

Disentangling the Systematic from the Merely Fortunate

While it was correct to forecast that commercial firms would restructure Internet access to suit commercial users, many users did not need such restructuring to make use of the technology. Internet access obtained widespread commercial appeal without restructuring of operations and other facets of supply. As noted, this occurred for many reasons, but two historically unique factors heavily shaped the story. First, the Internet was a demonstrably viable network prior to its commercialization, already used by many researchers, a fact that aided its migration into commercial use through incremental change. Second, the invention of the web fueled commercial growth above and beyond what probably would have happened in any event. Will broad lessons emerge in spite of these particular circumstances?

Said another way, while it is better to be lucky than right, it is always better to be both right and lucky. Would the NSF have been right if they were unlucky? What if the browser had not been invented? Would we still be lauding the NSF for pursuing policies friendly to commercialization? In that spirit, this section briefly considers two counterfactual questions: (1) Would outcomes have been similar in the absence of the browser? (2) Would outcomes have been similar in the presence of the browser, but in the absence of NSF policies friendly to commercialization?

The Importance of the Browser   To answer a counterfactual question, it is important to ask: compared with what alternative set of events? This is difficult to answer in this instance because actual events had a certain inevitability to them. For example, consistent with its mandate as a public research institution, the University of Illinois encouraged diffusion of the browser through licensing (Cusumano and Yoffie 1998). To be sure, the Netscape browser of 1995 was a match thrown into a dry field, but parts of that field had already been set ablaze. After the University of Illinois began licensing Mosaic it was only a matter of time before the blaze became an inferno. In other words, if Netscape had not commercialized the technology somebody else would have done so soon. As another example, if Tim Berners-Lee had abandoned his project before completion, it appears that somebody else eventually would have invented something similar. Tim Berners-Lee's invention of hypertext (and then the World Wide Web) culminated decades of work associated with making computing easier to use, more networked and more visual instead of textual (Waldrop 2001).

Hence, the browser and hypertext appear to have a certain inevitability to them. In that light, the most conservative counterfactual is this: What if hypertext and the browser had been invented a few years later? Would the Internet have commercialized successfully?

The answer would appear to be yes, though events might not have been as dramatic. There are several reasons for that assessment. First, e-mail alone would have motivated considerable household adoption of Internet access even without the browser. E-mail was among the most popular uses for the Internet in its early years, and also popular were many of the community bulletin boards, financial applications, news, and chat rooms. There were some substitutes for these activities even without the Internet, but e-mail (especially) would have been difficult to recreate in private networks on a national level and would have compelled some commercial activity. Both households and businesses found this application useful and all surveys of Internet use place it as the most popular application (Clemente 1998, Nie and Ebring 2000). While some of the more visual applications in the bulletin board industry, such as commercial pornography and probably much electronic retailing, would not have moved to the Internet without the browser, surveys such as Clemente (1998) have never shown these as anything more than a fraction of early Internet use.

On a business level it is also possible to imagine considerable demand for Internet access even in the absence of the browser. Many of the same applications just discussed, such as e-mail and news, motivated business demand. In addition, much of the online database industry would have found benefit from moving to TCP/IP based file transfer as a substitute for bulletin board based file transfers that were more cumbersome for users than a standard FTP or telnet download. With some challenges to overcome, commercial transactions that were forced into EDI-based data transfers also would have found TCP/IP technology useful. However, it would have taken considerable time to shift many other database applications into this mode, so one should not underestimate the difficulties (which were considerable even with the browser). So it is reasonable to expect the growth of TCP/IP connections within private industry even without the browser, but not at such a high rate.

Even with a later invention of the browser, many of the other institutions supporting the development of the Internet also would still have been in place. The creation of the Internet Engineering Task Force would have continued to have an impact on standards development and diffusion. There might not have been anything similar to the World Wide Web Consortium, but the shareware movement would have continued, a factor that made it easier to obtain software for setting up independent ISPs. The computing industry had become sufficiently vertically disintegrated by the early 1990s to prevent any single firm from blockading diffusion of TCP/IP;[8] neither IBM's proprietary networking offerings, nor DEC's, nor anyone else's could have dominated networking communications standards the way TCP/IP did once it began to commercialize.

Finally, even without the browser, one would have expected some migration of online capabilities into commercial use at some level. Migration would not have been unusual by historical standards. New computing capabilities often incubate among technically sophisticated users, building up functionality over long periods of time before migrating into mainstream use (Bresnahan and Greenstein 1999). In this instance, the situation was ripe for migration. All the prototypes for text based online activity existed among sophisticated users. Moreover, the new functionality associated with Internet technologies did not require radical investments on the part of users to be commercially viable. To be sure, there was one historical novelty to the pattern of migration in this instance. Due to NSF restrictions on use, the sophisticated users of Internet access technology were primarily concentrated in research positions and at universities, a subset of sophisticated users in the computing industry. Aside from this feature, the broad pattern of incubation and migration resembles other episodes of platform and technological growth in computing.

This is not to take away credit from those who took the actions and made them happen, nor to deemphasize the importance of these events for firms, regions, and individuals. The contours of events most certainly would have played out differently if the browser had diffused later. It would have resulted in very different outcomes for particular companies, stockholders, and, arguably, regions where these companies locate. Without the browser subscription model, Internet access might have had lower adoption rates at businesses and homes, growth might not have been as explosive, and a different structure of supply might have arisen. However, it is important to recognize the broad pattern that arises irrespective of the contours of how it plays out: even without the browser Internet technology would have migrated into commercial markets and demand would arise under any scenario, motivating the industry to continue to grow to a substantial level.

The Importance of NSF Policies   Government employees deliberately let the baby bird out of the nest, encouraging its flight. NSF's policies enabled the entrepreneurial initiatives of commercial firms to influence migration of the technology. That said, migration of technology out of the research community into mainstream commercial markets might have happened under many government policies. So the question arises: Which government policies were critical? In light of later market events, the facet of NSF activities to highlight are those policies that did not turn exclusive use of the Internet into an idiosyncratic technology during its incubation.

There were many senses in which Internet technology was not isolated during its incubation. For example, after the NSF created the NSFNET in the mid 1980s there were no attempts to exclude researchers who had only mild research justifications for using the Internet, a policy decision that dated back to conflicts that arose when DARPA managed the precursor to the Internet. The diffusion of TCP/IP in the late 1980s further facilitated those goals, as it was an easy standard to use in virtually any computing network. The NSF also did not isolate the Internet from mainstream computing use or vendor supply, making contracts with firms such as IBM and MCI for operations, effectively subsidizing computing facilities at research facilities which did the same. In addition, the NSF developed and subsidized growth of the Internet at many locations, adopting a decentralized set of regional networks for its operation. This structure later facilitated private financing of Internet operations and further decentralization of investment decisions by organizations with commercial orientation.

It is possible to view other events in the late 1980s and early 1990s in a similar light. For one, NSF contracted with third parties, such as MERIT, for operations. These types of contracts prevented the network technology from being distant from mainstream engineering and technical standards. NSF permitted interconnection with private data communication firms, such as UUNet and PSI, a spin-off from one of the regional networks, well before commercial dial-up ISPs came into existence. These contracts also established precedents. Finally, NSF did not tightly police the use restriction, especially in the regional networks. Indeed, a number of staff worked toward a 1992 congressional law that officially lifted the use policy on NSFNET, providing more certainty that commerce could be conducted using assets that might have appeared (to a court) to be previously owned by the federal government.

Finally, it is important to note the absence of a particularly common error in large infrastructure development, the attitude of build it and they will come. That is, researchers and developers operating under government subsidies tend to fulfill their own vision of what to do with the technology instead of a user's. The NSF's actions effectively prevented this attitude from overwhelming development. As it turned out, the immediate use of Internet technology within academic research centers tended to put things to use quickly. It allowed researchers to find out what worked and why. Hence, some user desires influenced system design, operation, and growth—even prior to the emergence of organizations that have a commercial orientation and a direct incentive to take account of those desires.[9]

In summary, commercialization of government managed technology can fail because there is no incentive to anticipate technical, commercial, or structure challenges that may arise later in commercial markets. Since this failure can happen for a variety of reasons, it is not possible to point to any single NSF action as the policy that prevented such a failure or, alternatively, acted as the catalyst for commercialization. It is, however, accurate to say that the sum total of NSF's actions did not let exclusive use by researchers impose an irreversible idiosyncratic stamp on the Internet during formative periods of incubation. These policies did not generate an isolated technology nor foster creation of a nonmalleable operation around it. Instead, NSF incubated technology with features that could adapt to the demands of users who would later be in the majority. This is the broad goal worth emulating in policies for commercializing government managed technologies.

Challenges for the Near Future

The diffusion of broadband access, the widely forecast future for this industry, seems to be taking on a more typical pattern for new technology, where technical and commercial constraints shape the pattern of diffusion. It is unclear what the lowest cost method for the delivery of broadband services will be. It is also unclear what type of services will motivate mass adoption of costly high speed access to the home. There are technical limitations to retrofitting old cable systems and with developing DSL technology over long distances. It is unclear how many people will be willing to pay for such high speed services. These uncertainties cloud all forecasts. However, unlike in the past, there will not be 2 decades of incubation of broadband technology by only government sponsored researchers. Hence, there is no reason to anticipate anything like the speed of diffusion found in the dial-up market, nor take for granted that ubiquity will arise as easily (for more, see, e.g., Weinberg 1999, or Werbach 1997).

This observation would seem, at first blush, to suggest that this history sheds little light on the future—that past and future challenges are too unique to their time for comparison. However, that conclusion is a bit hasty. Looking forward in the ISP industry, it is possible to identify some technical, commercial, and structural challenges that resemble those of the past and that will alter the contours of behavior and outcomes. I will discuss some of these, cognizant that restructuring is still taking place and changing sufficiently fast, so that any discussion runs the risk of becoming obsolete as soon as it is written.

Lesson 1: The Past Does Offer Guidance for Understanding Patterns of Restructuring   The names of the firms may change and so too may the specifics of the strategies, but the absence of uniformity in the development of Internet access business models should persist into the future. New applications for web technology are still under development because the technology has potential beyond its present implementations. Not all local markets will experience the same type of competitive choices in access, nor should they. Not all vendors will see the same opportunities and these differences arise for sound economic reasons. Users with more experience still adopt applications closer to the frontier, while users with less experience still demand more refined applications. Web technology enables these differences to manifest in new directions and it is not obvious which implementation will succeed with either type of user. In other words, most of the economic fundamentals leading to structural challenges have not disappeared; hence, experimentation with new business models will probably continue.

Lesson 2: The Subscription Model of Internet Access Will Continue to Change   Commercial markets inherited an organizational form from their academic ancestors, modifying it slightly for initial use. There is no reason to presume that it will maintain the same operational structure under competitive pressure. Indeed, it is presently under competition from a variety of alternate business models which use dial-up access to subsidize another activity. There are already hints of these potential changes as some ISPs charge very little for access and make up for the lost revenue with other services, such as networking, hosting, or web design. AOL has successfully combined access with its walled garden of content and AT&T appears intent on pursuing a unique approach of combining content and access. Other recent innovative firms include Netzero.com, which is the most successful to date of many firms that have tried to provide access for free and garner revenue through sales of advertising. There are also many other such experiments altering the explicit definition of basic service, embedding it with more than e-mail, but also with games, streaming, linking, and so on, that has the effect of changing the pricing structure too. It is not crazy to predict that access, by itself, could become absorbed into a bundle of many other complementary commercial services, slowly fading as the standalone service that existed in the academic domain.

Lesson 3: The Economics of Internet Diffusion Lie Behind Much of the Digital Divide   Internet access diffused more easily to some users and in some locations. The margin between adoption and nonadoption has become popularly known as the digital divide. If some of these outcomes are understood as temporary results of a young diffusion process, then many of the differences between those with virtual experience and those without can be framed as the byproduct of the economic factors shaping this diffusion episode. Within business the important factors influencing adoption are the density of the location of the business, the availability of basic computer support services nearby, and a firm's previous investment in IT. At the home the important determinants are availability (which is influenced by density) as well as the same factors behind the diffusion of PCs: age, education, and income especially, and also race for some income levels. It follows, therefore, that policies aimed at digital divide issues, such as the E-rate program, should not address those factors which are only temporary and will resolve themselves through market forces without government intervention. Instead, government programs should target factors that are likely to be more durable over time and that lead to division in adoption behavior; such as density of location, income, education, and race.

Lesson 4: Geographic Pervasiveness Introduces New Economic Considerations   There is one additional reason to expect the typical business model to remain unsettled. Geographic pervasiveness has entered into calculations today and it was not a relevant consideration at the outset of commercialization. The pervasiveness of the Internet across the country (and the developed world) changes the economic incentives to build applications on top of the backbone, and alters the learning process associated with its commercial development. All ISPs now depend on each other at a daily level in terms of their network security, reliability, and some dimensions of performance. Many new applications—e.g., virtual private networking, voice telephony over long distances, multiuser conferencing, some forms of instant messaging, and gaming—require coordinating quality of services across providers.

It is still unclear whether new business models are needed to take advantage of applications that presume geographic pervasiveness. If so, firms with national backbones and assets will have a commercial advantage. Pervasiveness also changes the activities below the backbone in the vertical chain. It has altered the scale of the market for supplying goods and services to the access industry, altering the incentives of upstream suppliers, equipment manufacturers, or middleware software providers, to bring out new services and inventive designs for the entire network. This factor was also not present in the academic network and it is unclear how it will influence the structure of the industry moving forward.

Lesson 5: Is There a Need for New Communications Policy for the New Millennium?   Until recently, the place of technical change in most communications services was presumed to be slow and easily monitored from centralized administrative agencies at the state and federal level. It is well-known that such a presumption is dated, but it is unclear what conceptual paradigm should replace it. This paper illustrated how vexing the scope of the problem will be. In this instance, ISPs addressed a variety of commercial and structural challenges with little government interference, but under considerable technical and commercial uncertainty. This occurred because many legacy regulatory decisions had previously specified how commercial firms transact with the regulated public switch network. These legacy institutions acted in society's interest in this instance, fostering experimentation in technically intensive activities, enabling decentralized decision making to shape commercial restructuring in specific places and time periods. To put it simply, it was in society's interest to enhance the variety of approaches to new commercial opportunities and the existing set of regulations did just that. However, going forward it is unclear whether these legacy institutions are still appropriate for other basic staples of communications policies, such as whether a merger is in the public interest, whether incumbent cable firms should be mandated to provide open access, whether communications infrastructure should be subsidized in underserved areas, and whether Internet services should be classified as a special exemption, immune from taxation and other fiscal expenses. Hence, this industry is entering an era where market events and unceasing restructuring will place considerable tension on long-standing legal foundations and slow regulatory rule making procedures.

Notes

For the NBER program on Innovation Policy and the Economy. Associate Professor, Northwestern University and Research Associate, NBER. This paper cites liberally from research supported by the Council on Library Resources, the Kellogg Graduate School and the National Science Foundation. The author thanks Angelique Augereau, Oded Bizan, Tim Bresnahan, Barbara Dooley, Chris Forman, Amy Friedlander, Avi Goldfarb, Brian Kahin, Mort Rahimi, Scott Stern, Mitchell Waldrop and many seminar participants for comments. I am particularly grateful to Zvi Griliches who encouraged this research when it was at a formative stage. All remaining errors are mine alone.

1. The literature on general purpose technologies (Bresnahan and Trajtenberg 1995, Helpman 1998) also helps frame these themes by highlighting the role of coinvention, defined as the complementary inventions which make advances in general purpose technologies valuable for particular organizations in particular places at particular points in time.

2. For example, see studies of the supersonic transport (Cohen and Noll 1990), nuclear power (Cowan 1988), air frames (Mowery and Rosenberg 1992) and the early history of computing (Flamm 1989, Katz and Phillips 1982), among many such examples.

3. If anything, regulatory decisions for reciprocal compensation of competitive location exchange providers (CLECs) encouraged CLEC entry, which also partly encouraged ISP entry through interconnection with CLECs. Though important to incumbent local exchange carriers, however, one should not exaggerate this too much. The scale of this phenomenon grew tremendously in the late 1990s, but ISP entry started well before then. Moreover, since CLEC entry was primarily concentrated in dense urban areas, much of this effect was felt in urban areas, which would have experienced a great deal of ISP entry even without this implicit subsidy to CLECs.

4. For further documentation of these methods, see Downes and Greenstein 1999 or Greenstein 1999. The fall 1996 data covers over 14,000 phone numbers for over 3,200 ISPs. The fall 1998 data covers over 65,000 phone numbers for just under 6,000 ISPs.

5. Some ISPs have told me in interviews that this growth was initially in response to customer requests for local phone numbers for accessing networks (e-mail mostly at first) when these customers traveled outside their primary area. More recently, it is also common to have ISPs discuss the possibility of developing a large customer base for purposes of selling the base to a high bidder in some future industry consolidation.

6. Speed is the sole dimension for differentiating between frontier and basic access. This is a practical choice. There are a number of other access technologies just now becoming viable that are slow but technically difficult, such as wireless access. Only a small number of firms in this data offer these services and these firms also offer high speed access.

7. One of the most difficult phrases to classify was general "consulting." The vast majority of consulting activity is accounted for by the present classification methods as one of these three complementary activities, networking, hosting, and web design.

8. The direction of commercial events also would have continued to take the same directions. Important among them was the final dissolution of the working relationship between IBM and Microsoft, as well as the final triumph of Ethernet-based standards within the majority of networking equipment for LANs.

9. One might ask why NSF adopted these policies when they did and whether their consequences were anticipated. That is a longer story and beyond the scope of this paper,

which was simply to highlight those which were useful in light of later events. The account of Waldrop (2001), for example, begins that evaluation by arguing that NSF was making virtue out of necessity. He argues that there was no expectation that government agencies could operate a large scale data network indefinitely. This was particularly so at NSF, whose budget was periodically realigned by the whims and fads of political fashion. There also was no expectation that NSF could or would fund decades worth of large scale data communications research on the scale that DARPA had done. Hence, it was believed that a sustainable network would necessarily require private partnership on some level and, eventually, private financing.

References

Augereau, Angelique, and Shane Greenstein. 1999. "The Need for Speed in Emerging Communications Markets: Upgrades to Advanced Technology at Internet Service Providers." Available at http://www.kellogg.nwu.edu/faculty/greenstein/images/research.html.

Boardwatch Magazine. 1996–1998. *Directory of Internet Service Providers*. Littleton, CO.

Bresnahan, Timothy, and Shane Greenstein. 1999. "Technological Competition and the Structure of the Computing Industry." *Journal of Industrial Economics* 47:1–40.

Bresnahan, Timothy, and Manuel Trajtenberg. 1995. "General Purpose Technologies: Engines of Growth?" *Journal of Econometrics* 65:83–108.

Clemente, Peter C. 1998. *The State of the Net: The New Frontier*. New York: McGraw-Hill.

Cohen, Linda, and Roger Noll. 1990. *The Technology Pork Barrel*. Washington, DC: Brookings Institution.

Cowan, Robin. 1988. *Nuclear Power Reactors: A Study in Technological Lock-in*, Working Paper no. 88-33, C.V. Starr Center for Applied Economics, New York University, New York.

Cusumano, Michael, and David Yoffie. 1998. *Competing on Internet Time, Lessons from Netscape and Its Battle With Microsoft*. New York: The Free Press.

Demsetz, Harold. 1988. "The Theory of the Firm Revisited." *Journal of Law, Economics, and Organization* 4:159–178.

Downes, Tom, and Shane Greenstein. 1998. "Do Commercial ISPs Provide Universal Access?" in Sharon Gillett and Ingo Vogelsang, eds., *Competition, Regulation and Convergence: Selected Papers from the 1998 Telecommunications Policy Research Conference*. Mahwah, NJ: Lawrence Erlbaum Associates.

Flamm, Kenneth. 1989. *Targeting the Computer, Government Support and International Competition*. Washington, DC: Brookings Institute.

Greenstein, Shane. 1999. "Building and Developing the Virtual World: The Commercial Internet Access Market." Available at http://www.kellogg.nwu.edu/faculty/greenstein.

Griliches, Zvi. 1957. "Hybrid Corn: An Exploration in the Economics of Technological Change." *Econometrica* 25:501–22.

Helpman, Elhanan. 1998. *General Purpose Technologies and Economic Growth*. Cambridge, MA: MIT Press.

Kalakota, Ravi, and Andrew Whinston. 1996. *Frontiers of Electronic Commerce.* Reading, MA: Addison-Wesley.

Katz, Barbara, and Almiran Phillips. 1982. "The Computer Industry." In Richard R. Nelson, ed., *Government and Technical Progress: A Cross-Industry Analysis:* 162–232. New York: Pergamon Press.

Kolstad, Rob. 1998, January. "Becoming an ISP." Available at www.bsdi.com.

Maloff Group International, Inc. 1997, October. "1996–1997 Internet Access Providers Marketplace Analysis." Dexter, MO.

Mowery, David, and Nathan Rosenberg. 1992. *Technology and the Pursuit of Economic Growth.* New York: Cambridge University Press.

National Telecommunication and Information Administration. 1999, July. *Falling Through the Net: Defining the Digital Divide.* U.S. Department of Commerce.

Nelson, Richard, and Sidney Winter. 1977. *An Evolutionary Theory of Economic Change.* Cambridge, MA: Belknap Press.

Nie, Norman, and Lutz Ebring. 2000. *Internet and Society: A Preliminary Report.* Palo Alto, CA: Stanford Institute for the Quantitative Study of Society, Stanford University.

Sidek, Greg, and Dan Spulber. 1998. "Cyberjam: The Law and Economics of Internet Congestion of the Telephone Network." *Harvard Journal of Law and Public Policy* 21(2).

Strover, Sharon. 1999. "Rural Internet Connectivity." Working Paper, University of Texas at Austin.

Ventresca, Mark, Rodney Lacey, Michael Lounsbury, and Dara Szyliowicz. 1998. "Industries as Organizational Fields: Infrastructure and Formative Dynamics in U.S. Online Database Services." A Working Paper in the "Business and Government" Series of the Institute for Policy Research, Northwestern University.

Waldrop, Mitchell. 2001, forthcoming. *Technology of Enchantment.* New York: Viking Press.

Weinberg, Jonathan. 1999, Spring. "The Internet and Telecommunications Services, Access Charges, Universal Service Mechanisms, and Other Flotsam of the Regulatory System." *Yale Journal of Regulation.*

Werbach, Kevin. 1997, March. *A Digital Tornado: The Internet and Telecommunications Policy.* FCC, Office of Planning and Policy Working Paper no. 29.

6

Numbers, Quality, and Entry: How Has the Bayh-Dole Act Affected U.S. University Patenting and Licensing?

David C. Mowery, *University of California, Berkeley*
Arvids A. Ziedonis, *University of Pennsylvania*

Executive Summary

This paper summarizes the results of empirical analyses of data on the characteristics of the pre- and post-1980 patents of three leading U.S. academic patenters—the University of California, Stanford University, and Columbia University. We complemented this analysis of these institutions with an analysis of the characteristics of the patents issued to all U.S. universities before and after 1980. Our analysis suggests that the effects of the Bayh-Dole Act on the content of academic research and patenting at Stanford and the University of California were modest. The most significant change in the content of research at these universities, one associated with increased patenting and licensing at both universities before and after 1980, was the rise of biomedical research and inventive activity, but Bayh-Dole had little to do with this growth. Indeed, the rise in biomedical research and inventions in both of these universities predates the passage of Bayh-Dole. Both UC and Stanford university administrators intensified their efforts to market faculty inventions in the wake of Bayh-Dole. This enlargement of the pool of marketed inventions appears to have reduced the average "yield" (defined as the share of license contracts yielding positive revenues) of this population at both universities. But we find no decline in the "importance" or "generality" of the post-1980 patents of these two universities. The analysis of overall U.S. university patenting suggests that the patents issued to institutions that entered into patenting and licensing after the effective date of the Bayh-Dole Act are indeed less important and less general than the patents issued before and after 1980 to U.S. universities with longer experience in patenting. Inexperienced academic patenters appear to have obtained patents that proved to be less significant (in terms of the rate and breadth of their subsequent citations) than those issuing to more experienced university patenters. Bayh-Dole's effects on entry therefore may be as important as any effects of the Act on the internal "research culture" of U.S. universities in explaining the widely remarked decline in the importance and generality of U.S. academic patents after 1980.

I. Introduction

The U.S. research university has played a central role in the evolution
of the U.S. innovation system during this century. Indeed, both the U.S.
research university and the organized pursuit of R&D in industry trace
their origins back roughly 125 years, and have grown in parallel
throughout the 20th century (Mowery and Rosenberg 1998). Although
links between R&D in U.S. industry and research in U.S. universities
have a long history, recent developments in this relationship, especially
the growth in university patenting and licensing of technologies to private firms, have attracted considerable attention.

U.S. university patenting and licensing have grown significantly in
the wake of an important federal policy initiative, known as the
Bayh-Dole Act of 1980. Although the Act's importance is widely cited,
its effects on U.S. research universities and on the U.S. innovation system have been the focus of little empirical analysis (Henderson et al.
1998 and Trajtenberg et al. 1997 are important exceptions). This paper
summarizes the results of empirical analyses of data on the characteristics of the pre- and post-1980 patents of three leading U.S. academic
patenters—the University of California (UC), Stanford University, and
Columbia University. We complemented this analysis of these institutions with an analysis of the characteristics of the patents issued to all
U.S. universities before and after 1980 (See Mowery et al. 1999 and
Mowery and Ziedonis 2000 for the detailed analyses).

We undertook this empirical research to assess the effects of
Bayh-Dole on patenting and licensing at two universities (UC and
Stanford) with substantial pre-1980 experience in these activities,
and to compare these universities with Columbia University, a post-
1980 entrant into patenting and licensing. This research examined the
widespread assertion that the increased patenting activity of U.S. universities during the 1980s was due to the Bayh-Dole Act. The second
piece of empirical research that is summarized in this paper examined
Bayh-Dole's effects on the content of academic research. We analyzed
Bayh-Dole's effects on UC and Stanford's technology marketing efforts, as well as the "importance" and "generality" of the patents issuing to these universities, before and after 1980. Finally, we compared
these characteristics of the patents issuing to these two universities
with those of Columbia University, and examined the characteristics of
patents issuing to all U.S. universities before and after 1980.

Our before and after analysis suggests that the effects of Bayh-Dole on the content of academic research and patenting at Stanford and the University of California were modest. The most significant change in the content of research at these universities, one associated with increased patenting and licensing at both universities before and after 1980, was the rise of biomedical research and inventive activity, but Bayh-Dole had little to do with this growth. Indeed, the growth in biomedical research and inventions in both of these universities predates the passage of Bayh-Dole. Both UC and Stanford university administrators intensified their efforts to market faculty inventions in the wake of Bayh-Dole, expanding the pool of university inventions for which patent applications were made and licensees sought. This enlargement of the pool of marketed inventions appears to have reduced the average yield (defined as the share of license contracts yielding positive revenues) of the intensified technology marketing efforts of both universities after 1980. But we find no decline in the importance or generality of the post-1980 patents of these two universities.

The analysis of overall U.S. university patenting suggests that the patents issued to institutions that entered into patenting and licensing after the effective date of the Bayh-Dole Act are indeed less important and less general than the patents issued before and after 1980 to U.S. universities with longer experience in patenting. In other words, an important factor in any assessment of Bayh-Dole's effects on U.S. academic patenting is the entry by universities with little experience into patenting and licensing after 1980. These inexperienced academic patenters appear to have obtained patents that proved to be less significant (in terms of the number and breadth of their subsequent citations) than those issuing to more experienced university patenters. Bayh-Dole's effects on entry therefore are at least as important as any effects on the internal research culture of U.S. universities in explaining the widely remarked decline in the importance and generality of U.S. academic patents after 1980 (See Henderson et al. 1998).

Immediately below, we discuss the background to the Bayh-Dole Act, in order to highlight the point that university-industry linkages, university patenting, and university licensing of these patents are not new features of the U.S. innovation system. We then discuss our data for these universities, present the comparative analysis, and consider the implications of our findings for the academic research enterprise and the overall innovation system of the United States.

II. Historical Background

The historic involvement of publicly funded universities in the United States with agricultural research, much of which was applied in character, and the involvement of these universities with the agricultural users of this research, are well-known aspects of U.S. economic history.[1] But throughout this century, the decentralized structure of U.S. higher education and the dependence of public and private universities on local sources of funding also meant that in a broad array of nonagricultural fields, ranging from engineering to physics and chemistry, collaborative research relationships between university faculty and industry were common (Rosenberg and Nelson 1994).

World War II transformed the role of U.S. universities as research performers and the sources of U.S. universities' research funding. The share of industry funding declined within the expanded research budgets of postwar U.S. research universities during the 1950s and 1960s, and by the early 1970s, federal funds accounted for more than 70% of university-performed research and industrial funding accounted for 2.6%. Much university research nevertheless retained an applied character, reflecting the importance of research support from such federal mission agencies as the Defense Department.

Beginning in the 1970s, the share of industry funding within academic research began to grow again. By 1997, federal funds accounted for 59% of total university research, and industry's share of the overall U.S. university research budget had tripled to more than 7% (all data are from National Science Board, 1996, and National Science Foundation, 2000). Most of the increase in industry funding occurred during the 1980s, and the industry share of university research funding has changed very little since 1990.

In view of the applied character of a good deal of their research, it is not surprising that a number of U.S. research universities were active in patenting and licensing faculty inventions well before 1980. Beginning in 1926, the University of California required all employees to report patentable inventions to the university administration. Other universities, like MIT and the University of Wisconsin, also developed administrative units to help patent and license inventions resulting from research. But relatively few academic institutions assumed direct responsibility for management of these activities, choosing instead to leave them in the hands of individual faculty or relying on external organizations such as the Research Corporation (Mowery and Sampat 1999).

Expanded federal research funding during the postwar period rekindled the debate over the disposition of the results of academic research (See Eisenberg 1996 for a review of the history of these policy debates). During the 1960s, both the Defense Department and the Department of Health, Education and Welfare (now HHS, the agency housing the National Institutes of Health), which were among the leading sources of federal academic research funding, allowed academic institutions to patent and license the results of their research under the terms of Institutional Patent Agreements (IPAs) negotiated by individual universities with each federal funding agency. IPAs eliminated the need for case-by-case reviews of the disposition of individual academic inventions and facilitated licensing of such inventions on an exclusive or nonexclusive basis, but tensions between some major IPA participants, such as the University of California, and their federal research sponsors remained.[2] These debates intensified in the late 1970s, when HEW began to question the use by some U.S. universities of exclusive licenses under IPAs, and proposed limiting the ability of some universities to adopt such policies.

The Bayh-Dole Patent and Trademark Amendments Act of 1980 provided blanket permission for performers of federally funded research to file for patents on the results of such research and to grant licenses for these patents, including exclusive licenses, to other parties. The Act facilitated university patenting and licensing in at least two ways. First, it replaced the web of IPAs that had been negotiated between individual universities and federal agencies with a uniform policy. Second, the Act's provisions expressed Congressional support for the negotiation of exclusive licenses between universities and industrial firms for the results of federally funded research.

The passage of the Bayh-Dole Act was one part of a broader shift in U.S. policy toward stronger intellectual property rights.[3] Among the most important of these policy initiatives was the establishment of the Court of Appeals for the Federal Circuit (CAFC) in 1982. Established to serve as the court of final appeal for patent cases throughout the federal judiciary, the CAFC soon emerged as a strong champion of patentholder rights.[4] But even before the establishment of the CAFC, the 1980 U.S. Supreme Court decision in *Diamond v. Chakrabarty* upheld the validity of a broad patent in the new industry of biotechnology, facilitating the patenting and licensing of inventions in this sector. The effects of Bayh-Dole thus must be viewed in the context of this larger shift in U.S. policy toward intellectual property rights.

The period following the passage of the Bayh-Dole Act was characterized by a sharp increase in U.S. university patenting and licensing activity. The data in table 6.1 reveal a large increase in university patenting after 1980—the number of patents issued to the 100 leading U.S. research universities (measured in terms of their 1993 R&D funding) more than doubled between 1979 and 1984, and more than doubled again between 1984 and 1989. Henderson, Jaffe, and Trajtenberg (1994) noted that the share of all U.S. patents accounted for by universities grew from less than 1% in 1975 to almost 2.5% in 1990. Moreover, the ratio of patents to R&D spending within universities almost doubled during 1975–1990 (from 57 patents per $1 billion in constant-dollar R&D spending in 1975 to 96 in 1990), while the same indicator for all U.S. patenting displayed a sharp decline (decreasing from 780 in 1975 to 429 in 1990). In other words, universities increased their patenting per R&D dollar during a period in which overall patenting per R&D dollar was declining.

In tandem with increased patenting, U.S. universities expanded their efforts to license these patents. The Association of University Technology Managers (AUTM) reported that the number of universities with technology licensing and transfer offices increased from 25 in 1980 to 200 in 1990, and licensing revenues of the AUTM universities increased from $183 million to $318 million in the 3 years from 1991 to 1994 alone (Cohen et al. 1998). As these data suggest, the Bayh-Dole Act triggered the entry by many universities into patenting and licensing. But even at incumbent academic patenters and licensors, the 1980s were marked by intensified technology licensing activity.

Table 6.1
Number of U.S. Patents Issued to 100 U.S. Academic Institutions with the Highest 1993 R&D Funding , 1974–1994

Year	Number of U.S. Patents
1974	177
1979	196
1984	408
1989	1004
1994	1486

National Science Board, 1996.

III. The Effects of Bayh-Dole on the Content of UC and Stanford Disclosures, Patenting, and Licensing

Bayh-Dole and the Rise of Biomedical Research at UC and Stanford University, 1975–1988

Both the University of California system and Stanford University established offices to promote the patenting and licensing of faculty inventions well before the passage of the Bayh-Dole Act. In 1963, the UC Board of Regents adopted a policy stating that all "Members of the faculties and employees shall make appropriate reports of any inventions and licenses they have conceived or developed to the Board of Patents."[5] In 1976, responsibility for patent policy was transferred from the General Counsel to the Office of the UC President, and in 1991 the Patent Office was reorganized into the Patent, Trademark, and Copyright Office (PTCO), and renamed the Office of Technology Transfer (OTT).

Stanford University's Office of Technology Licensing was established in 1970, and Stanford was active in patenting and licensing throughout the 1970s. Disclosure by faculty of inventions and their management by Stanford's OTL was optional for most of OTL's first quarter-century, but in 1994 Stanford changed its policy toward faculty inventions in two important aspects. First, assignment of title to the University of inventions ". . . developed using University resources . . ." was made mandatory. Second, the University established a policy under which "Copyright to software developed for University purposes in the course of employment, or as part of either a sponsored project or an unsponsored project specifically supported by University funds, belongs to the University." ("Office of Technology Licensing Guidelines for Software Distribution," 11/17/94).[6] Before and after 1994, the Stanford data contain many more faculty software inventions than do the UC data, reflecting the fact that the mandatory disclosure policy of the University of California did not cover software, deemed at the time to be copyrightable rather than patentable intellectual property.

Since both of these universities were active in patenting and licensing well before the passage of the Bayh-Dole Act, a comparison of the 1975–1979 period (prior to Bayh-Dole) and 1984–1988, following the passage of the bill, provides a "before and after" test of the Act's effects. The average annual number of invention disclosures at the University of California during 1984–1988, following passage of the Bayh-Dole Act, is almost 237, well above their average level (140 annual disclosures) for the 1975–1979 period. The period following the Bayh-Dole Act thus is associated with a higher average level of annual invention disclosures (confirmed in figure 6.1); but the timing of the increase in annual disclosures suggests that more than the Bayh-Dole Act affected this shift.

Figure 6.2 displays a 3-year moving average for annual invention disclosures by UC research personnel, omitting the first and last years in the 1975–1988 period. Both figures indicate that the increase in the average annual number of invention disclosures predates the passage of the Bayh-Dole Act; indeed, the largest single year-to-year percentage increase in disclosures during the entire 1974–1988 period occurred in 1978–1979, before the Act's passage. This increase in disclosures may reflect the important advances in biotechnology that occurred at UC San Francisco during the 1970s, or other changes in the structure and activities of the UC patent licensing office that were unrelated to Bayh-Dole. For example, the Cohen-Boyer DNA splicing technique, the basis for the single most profitable invention licensed by the UC system and Stanford University, was disclosed in 1974 and the first of several patent applications for the invention was filed in 1978, well before the passage of Bayh-Dole (this patent was issued in 1980).

Since biomedical inventions account for the lion's share of UC patenting and licensing after 1980, our assessment of trends before and after Bayh-Dole focuses on biomedical inventions, patents, and licenses.

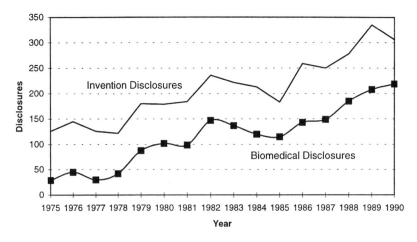

Figure 6.1
UC Invention Disclosures, 1975–1990

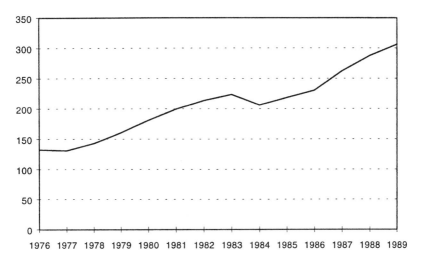

Figure 6.2
UC Invention Disclosures (Three Year Moving Average)

Figure 6.3 reveals that the shares of biomedical inventions within all UC invention disclosures began to grow in the mid-1970s, before the passage of Bayh-Dole. Moreover, these biomedical inventions accounted for a disproportionate share of the patenting and licensing activities of the University of California during this period: biomedical invention disclosures made up 33% of all UC disclosures during 1975–1979 and 60% of patents issued to the University of California for inventions disclosed during that period.[7] Biomedical patents accounted for 70% of the licensed patents in this cohort of disclosures, and biomedical inventions accounted for 59% of the UC licenses in this cohort that generated positive royalties. Biomedical inventions retained their importance during the 1984–1988 period, as they accounted for 60% of disclosures, 65% of patents, 74% of the licensed patents from this cohort of disclosures, and 73% of the positive-income licenses for this cohort of disclosures.

Data from the Stanford OTL provide similar "before and after" information on the patenting and licensing activities of a major private research university. Like those for the University of California, these data suggest that the growth of Stanford's patenting and licensing activities was affected by shifts in the academic research agenda that reflected influences other than Bayh-Dole. Figures 6.4 and 6.5 display trends during 1975–1990 in Stanford invention disclosures. The average annual number of disclosures to Stanford's Office of Technology

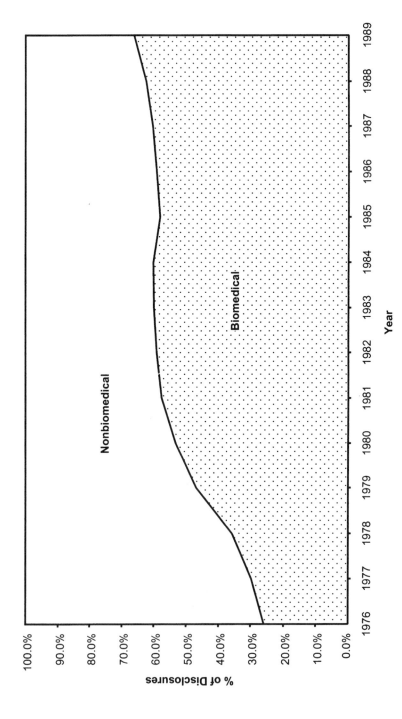

Figure 6.3
UC Biomedical Disclosures as a Percent of Total Disclosures, 1975–1990 (Three Year Moving Average)

Licensing increased from 74 during 1975–1979, prior to Bayh-Dole, to 149 during 1984–1988. The evidence of a "Bayh-Dole effect" on the annual number of disclosures (such as the jump in disclosures between 1979 and 1980 in figure 6.4) is stronger in the Stanford data than in the UC data.

The data in figures 6.4 and 6.6 suggest that the importance of biomedical inventions within Stanford's invention portfolio advances had begun to expand before the passage of Bayh-Dole. Figure 6.4 indicates that the annual number of biomedical invention disclosures began to increase sharply during the 1978–1980 period, and the share of all disclosures accounted for by biomedical inventions (see figures 6.4 and 6.6) increased steadily from 1977 to 1980, leveling off after 1980 and declining after 1983. The magnitude of these increases in biomedical inventions prior to Bayh-Dole is more modest than at the University of California, but the trend is similar. Biomedical inventions also increased their share of Stanford's (nonsoftware) licenses during the 1975–1990 period, although the upward trend is less pronounced and fluctuates more widely than in the UC data.

Additional evidence on the shifting composition of the UC and Stanford technology licensing portfolios is displayed in table 6.2. The UC

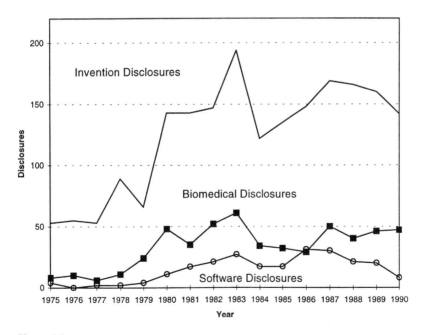

Figure 6.4
Stanford University Invention Disclosure, 1975–1990

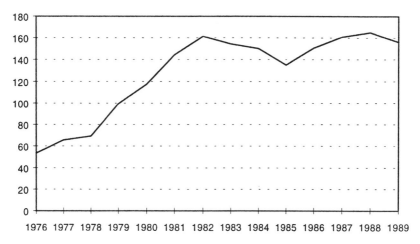

Figure 6.5
Stanford University 1975–1990 Invention Disclosures (Three Year Moving Average)

data in table 6.2 reveal the high concentration of licensing revenues among a small number of inventions throughout the pre-Bayh-Dole period, as well as indicating significant growth (more than 50-fold) in constant-dollar gross revenues during 1970–1995. Equally remarkable is the shift in the UC system's top 5 inventions from agricultural inventions (including plant varieties and agricultural machinery) to biomedical inventions. Among the three universities discussed in detail in this paper, only the University of California maintained a large-scale agricultural research effort. During the 1970s, agricultural inventions accounted for a majority of the income accruing to the top 5 UC money earners. Beginning in fiscal 1980, however, this share began to decline, and by fiscal 1995, 100% of the UC system's licensing income from its top 5 inventions, accounting for almost $40 million in revenues (in 1992 dollars), was derived from biomedical inventions, up from 20% in fiscal 1975. Moreover, and consistent with the previous discussion, this share increased before the passage of Bayh-Dole in late 1980: the share of the top 5 licensing revenues associated with biomedical inventions jumped from less than 20% in fiscal 1975 to more than 50% in fiscal 1980.

The data in table 6.2 on Stanford's licensing income display similar trends to those observed at UC. As of fiscal 1980, slightly more than 40% of the income from Stanford's top 5 inventions was attributable to biomedical inventions, suggesting the considerable importance of these inventions prior to Bayh-Dole. This share increased to more than

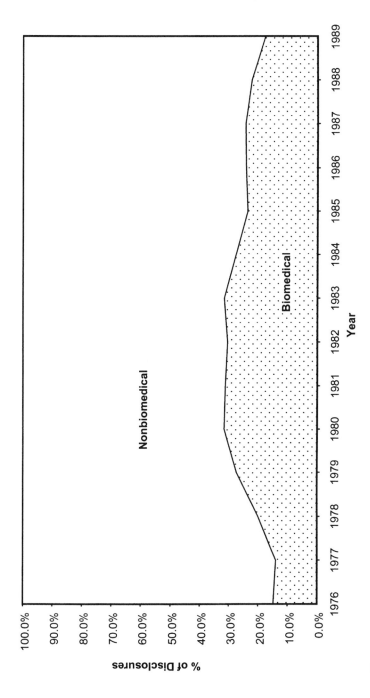

Figure 6.6
Stanford University Biomedical Disclosures as a Percent of Total Disclosures, 1975–1990 (Three Year Moving Average)

Table 6.2
Seclected Data on University of California, Stanford University, and Columbia University Licensing Income, FY1970–1995

UC	FY1970	FY1975	FY1980	FY1985	FY1990	FY1995
Gross income (1992 dollars: 000s)	1,140.4	1,470.7	2,113.9	3,914.3	13,240.4	58,556.0
Gross income from top five earners (1992 dollars: 000s)	899.9	1,074.8	1,083.0	1,855.0	7,229.8	38,665.6
share of gross income from top five earners (%)	79	73	51	47	55	66
share of income of top five earners associated with biomedical inventions (%)	34	19	54	40	91	100
share of income of top five earners associated with agricultural inventions (%)	57	70	46	60	9	0
Stanford		FY76				
Gross income (1992 dollars: 000s)	180.4	842.6	1,084.4	4,890.9	14,757.5	35,833.1
Gross income from top five earners (1992 dollars: 000s)		579.3	937.7	3,360.9	11,202.7	30,285.4
share of gross income from top five earners (%)		69	86	69	76	85
share of income of top five earners associated with biomedical inventions (%)		87	40	64	84	97
Columbia						
Gross income (1992 dollars: 000s)				542.0	6,903.5	31,790.3
Gross income from top five earners (1992 dollars: 000s)				535.6	6,366.7	29,935.8
share of gross income from top five earners (%)				99	92	94
share of income of top five earners associated with biomedical inventions (%)				81	87	91

96% by fiscal 1995. Stanford's licensing revenues grew almost 200-fold (in constant dollars) during 1970–1995, and its top 5 inventions account for a larger share of gross income for the 1980–1995 period than do the top 5 UC inventions.

Stanford and the UC system thus both experienced a shift in the composition of their invention and licensing portfolio in favor of bio-medical inventions prior to Bayh-Dole. This shift in the composition of the invention disclosures at both universities was a result in part of the large increases in federal funding of academic biomedical research, particularly research in molecular biology, that began in the 1960s and received an additional impetus with the Nixon Administration's "War on Cancer" of the early 1970s. The shift in the composition of invention disclosures also reflected the dramatic advances in scientific under-standing during the 1970s that laid the foundations for the biotechnol-ogy industry. Finally, the ability to obtain patents on and license the results of this research was strengthened by the judicial decisions and other shifts in federal policy that occurred during the late 1970s and early 1980s. The apparent effects of Bayh-Dole on Stanford and UC in-vention disclosures, patenting, and licensing are confounded with these other influences on academic research and the feasibility of pat-enting and licensing. Bayh-Dole was an important, but not a determi-native, factor in the growth and changing composition of patenting and licensing activity at these institutions. Much of the increase in bio-medical patenting and licensing at both Stanford and UC during the late 1970s and 1980s would have occurred in the absence of the Bayh-Dole Act, reflecting these other influences.

The Yield of Patenting and Licensing at UC and Stanford, 1975–1979 and 1984–1988

Although both UC and Stanford generated significant numbers of dis-closures, patents, and licenses before the passage of Bayh-Dole, these universities' patents and licenses grew significantly after 1980. What ef-fects did Bayh-Dole have on these universities' technology marketing activities? To examine this issue, we compared data on the yield of the technology marketing activities of UC and Stanford University before and after 1980 (See Mowery and Ziedonis 2000).

The UC and Stanford data do not reveal significant shifts in the char-acter of the inventions disclosed to UC or Stanford administrators for possible patenting after Bayh-Dole's passage, which suggests little change in the character of faculty research disclosures after 1980. The

UC data indicate that university administrators sought patent protection for a broader share of the underlying population of disclosures after 1980—the share of invention disclosures generating patent applications increased from 24% during 1975–1979 to more than 31% during 1984–1988. Interestingly, the share of UC patent applications resulting in patents shrank between these two periods (from more than 62% in 1975–1979 to nearly 44% in 1984–1988), suggesting that this larger set of post-Bayh-Dole patent applications declined somewhat in patentability. The post-Bayh-Dole period of invention disclosures also is associated with more intensive licensing of the patents resulting from these disclosures, as the share of UC patents that were licensed grew from slightly more than 25% during 1975–1979 to more than 35% during 1984–1988. We lack comparable data for Stanford on patent applications, but the share of Stanford patents that were licensed remained almost unchanged, increasing from 62.7% for 1975–1979 to 63.7% for 1984–1988.

The more extensive licensing efforts of the post-1980 period appear to have produced a decline in the yield of these licenses at both UC and Stanford, measured as the share of licenses yielding positive royalties. At UC, the share of licenses yielding positive income declined sharply, from more than 84% during 1975–1979 to 47% during 1984–1988. When Stanford's software licenses are excluded (so as to increase the comparability of the UC and Stanford data), this share declines from 90% during 1975–1979 to 76% in 1984–1988. But these indicators of decline in the yield of the marketing efforts of the UC and Stanford technology licensing offices need not imply inefficient or economically irrational behavior—after all, it is the marginal, rather than the average, returns to licensing activities that are most important in evaluating the returns to these institutions' licensing activities. Moreover, this measure of the yield of these universities' licensing activities does not capture the size of the revenue streams associated with the average or the marginal license before and after Bayh-Dole.[8] Nonetheless, these data provide indirect evidence of intensified technology marketing activities at both universities in the wake of Bayh-Dole.

IV. The Importance and Generality of Academic Patents from Incumbents and Entrants, 1975–1992

The indicators of the yield or productivity of the technology licensing activities at UC and Stanford suggest that in the first decade after the passage of the Act, these incumbent universities' technology transfer efforts intensified, with a concomitant decline in their yield. But this evidence does not address an important issue raised in the work of Henderson et al. (1998), who suggest that these intensified post-Bayh-Dole efforts to market faculty inventions were associated with the issue to U.S. universities of patents that were less important and less general, based on the patterns of citations to these patents. In order to examine this issue in more detail, we supplemented our data from UC and Stanford with information on the patenting and licensing activities of Columbia University, a leading post-Bayh-Dole entrant.

We included Columbia University patents in our analysis of the post-Bayh-Dole era in order to support a more detailed comparison between the patents issuing to an important post-Bayh-Dole entrant with the post-Bayh-Dole patents of our incumbent university patenters. For much of the pre-1980 period, Columbia University had no formal policy for the patenting or administration of faculty inventions, beyond a statement prohibiting the patenting of medical inventions that was rescinded only in 1975. Columbia's patent policy was significantly altered after the passage of Bayh-Dole. The new policy, which took effect on July 1, 1981 (the effective date of Bayh-Dole), reserved patent rights for Columbia and shared royalties with the inventor and department. In 1984, a new policy statement clarified and codified the rules: the policy mandated that faculty members disclose to the University any potentially patentable inventions developed with university resources. In 1989, Columbia's policy on reserving rights to the University for faculty inventions created with University resources was extended to cover software. Inventions were to be disclosed to Columbia's technology transfer office, the Office of Science and Technology Development, which was founded in 1982.

Although we define Columbia as an entrant academic patenter, reflecting the fact that this university developed an active patenting and licensing policy only after 1980, in fact Columbia did accumulate a modest portfolio of fewer than 10 patents during the 1975–1980 period. Interestingly, despite Columbia's later entry into patenting and licensing, the data in table 6.2 on the characteristics of Columbia licensing income indicate that by the end of its first decade of licensing activities, Columbia was reaping considerable gross revenues from licensing, suggesting that this entrant was quite unusual. Despite its status as an entrant into patenting and licensing, by the early 1980s Columbia University was (along with Stanford and the UC system) one of the leading U.S. academic recipients of licensing revenues.[9] The composition of Columbia University's licensing portfolio also was remarkably similar to those of UC and Stanford. In particular, Columbia's licensing revenues were highly concentrated among a small number of inventions, and this small group of home runs was dominated by biomedical inventions, just as was the case at UC and Stanford.

Comparing UC, Stanford, and Columbia University Patents

Our comparative analysis of the pre- and post-1980 patents from UC, Stanford, and Columbia Universities used patent citations to compute measures of the importance and generality of university patents. Patent citations have been used in numerous empirical analyses as measures of knowledge spillovers and other characteristics of firms' technological capabilities (Griliches 1990 provides an excellent survey of the strengths and weaknesses of patent-based measures). When the U.S. Patent and Trademark Office grants a patent, the granting officer includes a list of all previous patents on which the granted patent is based. This list is made public as part of the publication of the patent at the time it issues. The patent officer is aided in compiling a list of previous patents by the patent applicant, who is legally bound to provide with the application a list of all patents that constitute relevant prior art.[10] Citations of prior patents thus serve as an indicator of the technological lineage of new patents, much as bibliographic citations indicate the intellectual lineage of academic research.

Our before and after analysis of UC and Stanford patents used the year in which the invention was first disclosed as the key datum in categorizing faculty disclosures and any associated patents as falling into the pre-Bayh-Dole (before or during 1980) or post-Bayh-Dole (1981 or later) eras. We focused on forward citations in our analysis of changes in the importance and generality of UC, Stanford, and Columbia patents; that is, the number of citations received by each patent following its issue. Citations to patents typically peak 4–5 years after the date of issue of the cited patent, and data on citations to more recently issued patents therefore are right-truncated; that is, more recent patents are underrepresented in the citations data. In order to address this problem, our dataset included only citations to patents that occurred within 6 years of the year of issue of the patent, and our sample included only patents issued between 1975 and 1992. Our dataset also included a control sample of nonacademic patents for each of these three universities, spanning the same time period and replicating the distribution of the UC, Stanford, and Columbia patents among patent classes.[11] Our patent-citations data for all three universities were separated into biomedical and nonbiomedical classes, because of the importance of biomedical patents in the licensing activities of all three universities before and after 1980.

We used the number of citations to a patent during the 6 years following its issue as a measure of the importance of the patent, based on the assumption that citations form an index of sorts of the influence over subsequent inventive activity of the cited patent. Our comparison of citations to this sample of academic and nonacademic patents yielded several interesting findings (see table 6.3 for a descriptive summary of the statistical results). All three universities' academic patents were cited more frequently than our control samples of nonacademic patents throughout the 1975–1992 period, suggesting a higher level of importance for the academic patents. Of greater importance for the issue being analyzed in this section is the lack of evidence of a decline in the relative importance of Stanford and UC patents, relative to nonacademic patents, in the post-Bayh-Dole period. These differences in relative citation rates for academic and nonacademic patents after 1980 were statistically significant overall, although the higher citation rates for Columbia's post-1980 patents were only statistically significant at the 10% level.

The results for these subsamples must be interpreted with considerable caution, since we have a relatively small number of observations for some time periods and technology fields. The relatively infrequent significant differences in importance between the university and control sample biomedical patents' citations is surprising, in view of the importance of biomedical patents within the patenting and licensing activities of Stanford and UC before and after Bayh-Dole. But the data provide no indication of any decline in the importance of these universities' patents, relative to our control samples of nonacademic patents, after the Bayh-Dole Act. If anything, the data suggest that the UC and Stanford patents' relative importance increased, rather than declined, after 1980.

Although these results provide some evidence that the patents applied for during the 1980s by Columbia, a university that did not patent significantly prior to Bayh-Dole, were less important, relative to all nonacademic patents, than those of Stanford and UC during this period, they do not suggest that Columbia's patents were significantly less important than those in its nonacademic control sample. The absence of significantly greater citation rates for Columbia patents could reflect a less selective approach to patenting during the early years of its licensing activities by Columbia University, an institution with little

Table 6.3
Summary of Differences in Mean "Importance" and "Generality" between UC, Stanford, and Columbia University Patents vs. Control Group Patents for Inventions Disclosed and Patented Before and After Bayh-Dole

Importance (*defined as* citations to academic patents–citations to nonacademic patents):

	Overall			Biomedical			Non-Biomedical		
	UC	Stanford	Columbia	UC	Stanford	Columbia	UC	Stanford	Columbia
Inventions Disclosed and Patents Issued 1975–1980	+	+**	NA	+	+	NA	+	+**	NA
Inventions Disclosed and Patents Issued 1981–1992	+**	+**	+*	+**	+**	+	+**	+**	+

Generality (*defined as* generality index for academic patents–generality index for nonacademic patents):

	Overall			Biomedical			Non-Biomedical		
	UC	Stanford	Columbia	UC	Stanford	Columbia	UC	Stanford	Columbia
Inventions Disclosed and Patents Issued 1975–1980	+	+**	NA	+	+**	NA	+	+**	NA
Inventions Disclosed and Patents Issued 1981–1992	+**	+**	+**	+*	+**	+*	+**	+**	+*

Definitions of Symbols:
+: Difference between academic and nonacademic index is positive.
−: Difference between academic and nonacademic index is negative.
** denotes difference in means between university and nonacademic control group patents is significant at 5% level.
* denotes difference in means between university and nonacademic control group patents is significant at 10% level.

experience in patenting, than UC or Stanford.[12] We return to this issue below.

We also tested for changes in the generality of UC, Stanford, and Columbia patents before and after Bayh-Dole. The more widely cited a patent outside of its home patent class, the greater its generality, and arguably, the more significant the advance in knowledge represented by the patent. Following Henderson et al. (1998), we compute generality as follows:

$$GENERAL_i = 1 - \sum_{k=1}^{N_i} \left(\frac{NCITING_{ik}}{NCITING_i} \right)^2$$

where for patent i, k is the index of patent classes and N_i the number of different classes to which the citing patents belong. Higher values of $GENERAL_i$ indicate that a patent is cited in a broader array of technology classes, which we take to indicate greater influence on subsequent inventive effort in diverse fields.

Overall, the mean generality measures for overall UC, Stanford, and Columbia patents were higher than those for their respective control sample patents, excepting only UC patents applied for and issued before 1981 (table 6.3). We found no evidence of a decline in generality (relative to the control sample of nonacademic patents) in the UC and Stanford patents in the post-Bayh-Dole era. Indeed, the differences in mean generality between the overall UC and Stanford patents and their respective control samples were statistically significant (at the 5% level) for the post-1980 period. The mean generality score for the post-1980 Columbia patents also was significantly higher (at the 5% level of significance) than was true of the patents in its control sample.

Overall, the results of this analysis of the importance and generality of Stanford, UC, and Columbia patents yield conclusions that differ somewhat from those of Henderson et al. (1998), who analyzed a larger sample of U.S. university patents. Why do we find little or no evidence of declines in the importance or generality of these universities' patents after the Bayh-Dole Act? First, our sample of university patents is small, although this should tend to favor findings of no statistically significant differences between the university and nonacademic patent samples. Second, these results ignore the effects of entry by other universities (other than Columbia) into patenting and licensing after 1980 on the characteristics of the overall U.S. academic patent portfolio in the post-Bayh-Dole era. Although Henderson et al. (1998) find similar evidence of declining importance in even the patents assigned to the 15 leading academic patenters (based on their 1988 patents) during the 1975–1988 period, this group also may have been affected by the entry of less experienced academic patenters. In the next section, we discuss the results of our analysis of the effects on the importance and generality of U.S. academic patents of entry by inexperienced academic patenters after 1980.

Entry, Importance, and Generality in Overall U.S. Academic Patenting, 1975–1991

We next analyzed a broader sample of all U.S. academic patents from the pre- and post-Bayh-Dole periods, in order to examine the effects on patent importance and generality of entry into patenting by inexperienced universities after 1980. These data allowed us to separate any change in the generality and importance of all U.S. academic patents after 1980 into those associated with incumbent universities' patents and those associated with the patents assigned to entrants, defined here as universities with few or no patent applications during our pre-Bayh-Dole period of 1975–1980.

The results of this analysis are intended to distinguish between two hypothesized effects of Bayh-Dole on academic research and patenting. Some observers have expressed concern that the expanded post-1980 efforts by U.S. universities to promote patenting and licensing of faculty inventions, especially when faculty share in the financial returns to these licenses, have skewed the content and character of university research to favor more applied research activities. Parallel declines in the importance and generality of the post-1980 patents of both incumbent and entrant universities indicate that the Bayh-Dole Act affected the incentives of academic researchers and (importantly) academic administrators to disclose and patent inventions of lower importance and generality throughout U.S. academia.

If these changes in the characteristics of U.S. post-1980 academic patents result from entry by less experienced patenters, however, a different interpretation of the effects of Bayh-Dole is plausible. For example, if new entrants initially patent a broad cross section of faculty discoveries, they may accumulate a patent portfolio of limited importance and generality (some anecdotal evidence supports this characterization). Over time, as they learn the complexities of protecting and marketing intellectual property and become more selective in their patenting, the gaps between the characteristics of their patents and those of the high-intensity incumbents could narrow somewhat, and any decline in the average generality or importance of overall U.S. academic patents should be attenuated.

The first interpretation of Bayh-Dole's effects emphasizes changed incentives and behavior throughout U.S. universities, while the other views the 1980s as a period of learning and adjustment to a new incentive environment by organizations inexperienced in patenting and licensing. Needless to say, these explanations are not mutually exclusive, and the development of U.S. academic patenting during the 1980s likely reflects both effects. But our empirical analysis should support an assessment of the relative strength of the two effects.

We probed the effects of entry by U.S. universities into patenting on the importance and generality of academic patents by constructing a dataset of all patents assigned to U.S. universities other than Stanford, UC, and Columbia during the 1975–1992 period. Within this dataset, we distinguished among three categories: (1) Universities with at least 10 patents that were applied for after 1970 and issued during 1975–1980; (2) Universities with at least one but fewer than 10 patents applied for after 1970 that were issued during 1975–1980; and (3) Universities with no patents issued during the 1975–1980 period and at least one patent issued during 1981–1992 that was applied for after 1980. Our definitions of entrant and incumbent universities are somewhat arbitrary, but we believe that this tripartite distinction enables us to separate the effects on patent importance and generality of increased patenting after Bayh-Dole by active pre-1980 patenters (a group that includes UC and Stanford) from increased post-1980 patenting by universities historically inactive in this area and increased patenting by universities (such as Columbia) that were minimally involved in patenting before 1980.

Figure 6.7 displays trends in the shares of all academic patents applied for after 1970 that are accounted for by these three groups during the 1975–1992 period. The figure illustrates the declining share of the high-intensity pre-1980 academic patenters after the passage of Bayh-Dole. The high-intensity patenters' share declines from more than 85% during 1975–1980 to less than 65% by 1992. The low-intensity pre-1980 patenters, by contrast, increase their share of all academic patents from 15% in 1981 to almost 30% in 1992. And entrants, institutions that display no patenting activity during 1975–1980, increase their share of overall academic patenting from zero in 1980 to more than 6% by 1992. The post-Bayh-Dole era clearly is characterized by a significant change in the population of academic patenters and a shift toward institutional patenters with less experience in this activity.

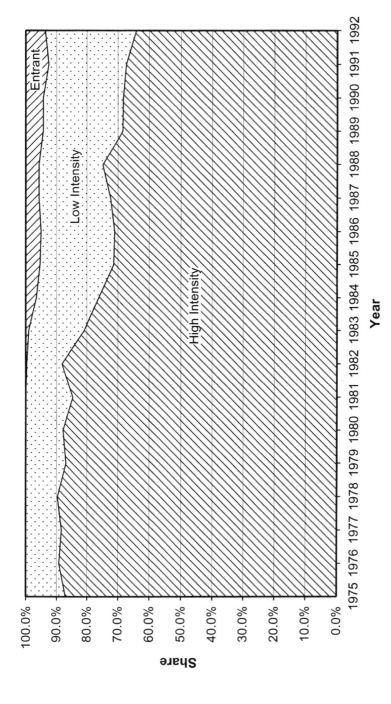

Figure 6.7
Shares of All University Patents by High Intensity, Low Intensity, and Entrant University Patenters, 1975–1992

We once again used patent citations data to analyze the importance and generality for the patents assigned to academic institutions in each of these three categories, covering 1975–1991 for the high- and low-intensity incumbents and 1981–1991 for the entrants.[13] Our statistical analysis covered overall academic patents, and separately examined biomedical and nonbiomedical patents. The results for our analysis of each of these three samples of academic patents display contrasting patterns of importance and generality. The patents assigned to the high-intensity incumbents are consistently more important and more general, relative to nonacademic patents, throughout the 1975–1991 period. Indeed, this group's patents were cited with increasing frequency after Bayh-Dole, relative to nonacademic patents.

The results for the other two groups of academic institutions whose patenting increased substantially after 1980 indicate consistently lower levels of importance and generality for their patents throughout the 1980s, relative to nonacademic patents. These results suggest that the patents of the two groups of U.S. universities that increased their share of overall academic patenting during the 1980s were of lower importance and generality, by comparison with a similar sample of nonacademic patents, than the patents issuing to U.S. universities with longer histories of patenting activity. The findings broadly corroborate our earlier finding of no decline in the importance and generality of post-1980 patents issuing to UC and Stanford University. Taken together, these results indicate that the deterioration in the importance and quality of post-1980 U.S. academic patents may have resulted from the Bayh-Dole-Act's encouragement of entry into patenting by academic institutions with relatively little experience in this activity.

In this view, Bayh-Dole's immediate effects on the content of academic research were modest, by comparison with the Act's encouragement of a new group of universities to expand or begin patenting of faculty inventions. With the passage of time, learning on the part of these entrants may gradually improve their management of patenting and licensing activities, and the apparent differences between their patent portfolios and those of the institutions long active in patenting may decline. We lack a sufficiently lengthy or rich longitudinal time series to test this possibility, although some anecdotal evidence is consistent with this characterization of many entrant academic institutions' patenting activities during the 1980s. But this interpretation of the effects of Bayh-Dole has different, and arguably less worrisome, implications for the future of the U.S. academic research enterprise than the alternative characterization noted above. The causes of any change during the

1980s in the characteristics of U.S. academic patents merit additional study.

V.   Conclusions and Policy Implications

Since the passage of the Bayh-Dole Act, many U.S. universities have expanded (or begun) programs to patent and license the results of federally and industrially funded research. Data from the University of California indicate that the Bayh-Dole Act was associated with an increase in the University's propensity to seek patent protection for faculty inventions that was not matched by a comparable increase in the yield of its patenting activities. At both UC and Stanford University, Bayh-Dole resulted in expanded efforts to market licenses to academic inventions. These expanded marketing efforts appear to have been associated with a modest decline in the yield of the invention marketing efforts at both of these institutions. At both universities, Bayh-Dole had modest effects on the content of academic research, since the composition of these institutions' invention portfolio had shifted before 1980 in favor of biomedical inventions that were highly attractive to commercial licensors. Moreover, the pre- and post-Bayh-Dole licensing efforts, and the revenue flows associated with these licensing efforts, tended to be concentrated in the biomedical area.

Nevertheless, the upsurge in patenting and licensing at these and other U.S. research universities after 1980 was affected by other factors in addition to the Bayh-Dole Act, and it is difficult to separate the effects of the Act from those of other factors. In particular, by the mid-1970s biomedical technology, especially biotechnology, had increased significantly in importance as a productive field of university research with research findings that were of great interest to industry. The feasibility of technology licensing in biotechnology was advanced by the *Diamond v. Chakrabarty* Supreme Court decision, which opened the door to patenting the organisms, molecules, and research techniques emerging from biotechnology. This judicial decision, as well as the broader shift in U.S. policy to strengthen intellectual property rights, contributed to the increased post-1980 patenting and licensing activities of U.S. research universities.[14]

This analysis of the effects of Bayh-Dole on the content of academic research and the importance of the patents assigned to these two leading research universities yields conclusions that both corroborate and contradict the findings of Henderson et al. (1998). The data on UC and Stanford invention disclosures before and after Bayh-Dole suggest some increase in the propensity of both institutions to patent and/or license faculty inventions. But our analysis of the importance of the pre- and post-Bayh-Dole patents assigned to these two universities reveals no systematic decline in importance or generality after Bayh-Dole, in contrast to the findings of Henderson et al. (1998). Nor do we find that the quality of patents accounted for by a major entrant, Columbia University, are significantly less important or general than those within a matched sample of nonacademic patents, although the Columbia patents are not significantly more important than those in the control sample for the post-1980 period.

How do we reconcile a finding that citation-based measures of UC and Stanford patents reveal no decline in importance after Bayh-Dole with our conclusion that both UC and Stanford's technology licensing operations appear to have experienced a decline in yield, that is, a decline in the share of licenses yielding positive revenues? Fundamentally, these two sets of indicators measure different characteristics of the invention and patent portfolios of these universities. Along with other scholars, we interpret patent citations as measures of the importance of the contribution to inventive knowledge of a given patent. But this contribution may or may not be correlated with the attractiveness to industry of a license for this patent. The extent of correlation between licensing revenues and patent citations is an important research question that we plan to examine in future work.

Our analysis of the effects of entry by less experienced academic patenters on the importance and generality of a much broader sample of post-1980 U.S. academic patents indicates that the patents of entrants with little or no previous history of patenting were not significantly more important or general than nonacademic patents. Given the significant expansion in the share of U.S. academic patents accounted for by these entrant institutions, it is plausible that the findings of Henderson et al. (1998) of declining importance and generality during the 1980s in U.S. academic patents reflect the effects of Bayh-Dole on entry, rather than on the incentives of academic researchers and administrators in long-active academic patenters. The evidence from the Columbia post-1980 patent sample, which is no less heavily or broadly cited than those of its nonacademic control sample, suggests that there is considerable heterogeneity within the population of entrant institutions.

The limitations of our analysis are apparent. We have detailed institutional data on patenting and licensing for only three very unusual universities, institutions that were among the leaders in patenting and licensing of faculty inventions before and after the Act's passage. The empirical results for both our three-university and broader academic samples are sensitive to the composition and construction of the patent control samples. In assessing the effects of Bayh-Dole on university technology transfer, we are analyzing only the formalized technology transfer activities of these universities, and cannot exclude the possibility that activities in invention disclosure, patenting, and licensing may affect the numerous other channels through which university knowledge reaches commercial application. Moreover, the small size of our samples of university patents, especially those covering the pre-Bayh-Dole period, limits the robustness and power of our statistical tests. Finally, this analysis of the post-Bayh-Dole period necessarily covers only the early years of this new regime. As denizens of any university can attest, change within these institutions occurs slowly, and it is possible that the true effects of the Bayh-Dole Act are only now being revealed. Nevertheless, we believe that the results of this analysis underscore the importance of complementing analyses of aggregate data on academic and nonacademic patenting trends with work on individual institutions engaged in these pursuits, be they firms, universities, or public laboratories.

What implications for policy emerge from this analysis? First and perhaps most important is the finding that more than Bayh-Dole alone underpinned the rise of U.S. universities' patenting and licensing activities after 1980. Without the rapid growth in federal biomedical research funding throughout the 1960–1980 period, as well as the other changes in federal policy toward intellectual property rights, the Bayh-Dole Act by itself would have had much less effect on university patenting and licensing. Indeed, this conclusion underscores a point made in the introduction to this paper—the post-1980 surge in patenting and licensing by U.S. universities is only the latest chapter in a long history of close ties between U.S. academic and industrial research.

The unusual institutional structure of the U.S. university system contributed to the strength and long history of such links. The large scale, high levels of institutional autonomy, and diversified source of public and private funding that characterize the U.S. higher education system have long created powerful incentives for faculty and administrators to seek external sources of research support, be these from private firms during the 1920s and 1930s, the Defense Department during the 1950s and 1960s, or industry during the 1980s and 1990s. Among other things, the importance of these other structural factors suggests that emulation of Bayh-Dole in other industrial economies with systems of higher education that are very differently structured may be counter-productive or unsuccessful.

Another important point from this discussion concerns the dominance of patenting and especially, licensing, by biomedical inventions at the three leading U.S. universities we analyzed in detail. There are several reasons for this dominance—patents in biomedical technologies are strong and very difficult to invent around; the field is characterized by an unusually close link between basic scientific research and commercial applications; and successful commercial applications are extremely profitable. Patenting and licensing thus may serve as reasonably effective channels for the transfer of academic research to commercial application in this technology. But in other fields of research, such as many areas of engineering, the strength of patents and the value of licenses are much lower.

University administrators and policymakers alike must recognize that universities transfer knowledge to commercial applications through a diverse array of channels, including the training of students, publication, faculty consulting, faculty involvement in new business enterprises, and more informal interactions, in addition to the licensing of patents. Moreover, the relative importance of these different channels varies among fields of research. Policies that aspire to uniformity across fields, especially where these policies borrow from the biomedical fields, may actually undercut the effectiveness of knowledge transfer and may reduce university-industry collaboration.

Although we find little evidence of significant change in the content of U.S. academic research in the wake of Bayh-Dole, and believe that many of the post-1980 changes that have occurred in the relationship between U.S. universities and industry would have occurred without Bayh-Dole, this does not mean that there are no reasons for concern over the effects of these new relationships on the academic research and teaching enterprise. There is an abundant supply of anecdotes of faculty conflicts of interest and universities that now include faculty patenting activities in their reviews of research excellence. Formal restrictions on publication or release of research results, or the informal discouragement of collaboration among faculty or students that may result from the growing commercial value of some academic research activities, pose real risks to graduate education in particular. In some fields of research, universities now are competing with industrial firms as much as they are collaborating—research tools, which many universities seek to patent and license to pharmaceutical and other biomedical firms, is one example. Strategic alliances between universities and individual firms, such as the agreement between UC Berkeley and the Novartis Corporation, are another example. Many of these challenges (such as conflicts of interest or restrictions on publication) are neither new nor attributable in their recent manifestations to the Bayh-Dole Act. They merit close scrutiny nonetheless.

Notes

Authors' names appear in alphabetical order. Earlier versions of this paper were presented at the 1999 AEA meetings and the July 1999 meeting of the NBER's Science and Technology Policy Group, and benefited from comments by participants at both meetings. Portions of this paper draw on an earlier paper coauthored with Richard Nelson and Bhaven Sampat of Columbia University (Mowery et al. 1999), and we have benefited from numerous conversations and comments from them on the issues covered in this paper. The paper also benefited from the comments of Josh Lerner, Adam Jaffe, and Scott Stern. We are indebted to the staff of the technology licensing offices of Stanford University, the University of California, and Columbia University for invaluable assistance with the collection and analysis of these data. Michael Barnes and Lynn Fissell of the University of California assisted in the collection and analysis of the University of California data, and the research on the Stanford data benefited from the assistance of Sandra Bradford. Special thanks to Michael Barnes for the use of his university patenting data and to Adam Jaffe of Brandeis University and the NBER for making his patent data available to us. Support for this research was provided by the California Policy Seminar, the U.C. President's Industry-University Cooperative Research Program, the Alfred P. Sloan Foundation, the Huntsman Center for Global Research of the Wharton School at the University of Pennsylvania, and the Andrew Mellon Foundation.

1. This section draws on Mowery et al. (1999).

2. According to the "Report on University Patent Fund and University Patent Operations for the Year Ended June 30, 1968" of the Board of Regents of the University of California, "The United States Public Health Service (PHS) of the Department of Health, Education, and Welfare is revising its Institutional Agreements under which patent rights can be retained by educational institutions. The PHS intends to make these Institutional Agreements available to many more institutions than at present. At the same time, it is making its patent provisions more restrictive. Most objectionable of the provisions included in the draft under consideration are: (1) a limitation on the amount of royalty the University can share with its inventors, and (2) a requirement that the University and its licensees provide the Government with copies of all licenses, and that the University incorporate into commercial licenses the provisions of the Institutional Agreement." (11/1/68, p. 4).

3. According to Katz and Ordover (1990), at least 14 Congressional bills passed during the 1980s focused on strengthening domestic and international protection for intellectual property rights, and the Court of Appeals for the Federal Circuit created in 1982 has upheld patent rights in roughly 80% of the cases argued before it, a considerable increase from the pre-1982 rate of 30% for the Federal bench.

4. See Hall and Ziedonis (2000) for an analysis of the effects of the CAFC and related policy shifts on patenting in the U.S. semiconductor industry.

5. The Board was a committee of UC faculty and administrators charged with oversight of the Patent Office. As revised in 1973, the "University Policy Regarding Patents" states that "An agreement to assign inventions and patents to The Regents of the University of California, except those resulting from permissible consulting activities without use of University facilities, shall be mandatory for all employees, academic and nonacademic." The policy statement goes on to emphasize that "The Regents is [sic] averse to seeking protective patents and will not seek such patents unless the discoverer or inventor can demonstrate that the securing of the patent is important to the University." This latter sentiment notwithstanding, UC administrators were actively seeking patent protection for faculty inventions by the mid-1970s, as the historical data of the Office of Technology Transfer show.

6. Reflecting faculty sensitivity over assignment to the University of all ownership of all copyrighted material produced under University sponsorship, Stanford's OTL explicitly exempted ownership of ". . . dissertations, papers, and articles, . . . popular nonfiction, novels, poems, musical compositions, or other works of artistic imagination which are not institutional works" from the policy governing software ("Copyrightable Works and Licensing at Stanford," Stanford University Office of Technology Licensing, Spring, 1994, p. 1).

7. Biomedical inventions accounted for a growing share of UC patenting and licensing during the entire 1975–1990 period.

8. For example, average income per license may have increased in the second period, although the skewed distribution of the licensing income of both the Stanford and UC technology transfer offices means that any such changes are likely to be small.

9. According to a recent report of the Association of University Technology Managers (AUTM) on institutional licensing income, fiscal 1997 gross licensing revenues for the UC system, Stanford University, and Columbia University amounted to $67.3 million, $51.8 million, and $50.3 million respectively. These three institutions ranked as the top three U.S. academic recipients of licensing income (Association of University Technology Managers 1997).

10. In addition to the legal requirement, it is in the applicant's interest to be forthcoming in this list because a more complete description of prior art is likely to reduce the prospects of an interference being declared during processing of a patent application.

11. Although our analysis followed that of Henderson et al. 1998 and Trajtenberg et al. 1997 closely, we employed a slightly different control population of patents, one that excludes academic patents and matches the distribution of our academic patent samples among patent classes.

12. Any such effect was significant during only the early years of Columbia's patenting and licensing activities, since by 1986–1990, the share of disclosures resulting in issued patents and the share of disclosures that result in licenses yielding positive royalty income are fairly similar at Columbia, UC, and Stanford (table 6.2).

13. Our analysis of the relative importance and generality of the patents of these three groups once again compared the patents from each group of universities with a control sample of patents constructed to replicate the distribution of the academic patents across time and among technology classes. Our regression analysis of importance used a negative binomial specification. We used a tobit specification in our regression analyses of generality, since this variable's distribution is truncated at a lower limit of zero and an upper limit of one. Each specification was estimated for a dataset covering the patents of

the relevant academic institutions and those in a matched control sample of nonacademic patents (matched by application year and patent class). We controlled for year effects and interacted a dummy variable denoting academic patents with a dummy variable for the application year.

14. But the influence of Bayh-Dole and broader changes in U.S. intellectual property rights policies, most of which affected patent coverage and enforcement, on academic technology licensing may be overstated. Stanford University was active before and after 1980, and Columbia became active after 1980, in licensing unpatented software inventions. For this technology class, the establishment of a university technology licensing office, rather than Bayh-Dole or other changes in U.S. patent policy, appears to have encouraged the disclosure and licensing of inventions whose intellectual property protection and ease of licensure were not affected by the Bayh-Dole Act.

References

Arrow, K. 1962. "Economic Welfare and the Allocation of Resources for Invention." In R. R. Nelson, ed., *The Rate and Direction of Inventive Activity.* Princeton, NJ: Princeton University Press.

Association of University Technology Managers. 1997. *AUTM Licensing Survey: Fiscal Year 1997.* Norwalk, CT: Association of University Technology Managers.

Bok, D. 1982. *Beyond the Ivory Tower.* Cambridge, MA: Harvard University Press.

Caves, R., Crookell, H., and P. Killing. 1983. "The Imperfect Market for Technology Licenses." *Oxford Bulletin of Economics and Statistics:* 249–67.

Cohen, W., R. Florida, and R. Goe. 1994. "University-Industry Research Centers in the United States." Pittsburgh, PA: Center for Economic Development, Carnegie-Mellon University. Technical report.

Cohen, W., R. Florida, L. Randazzese, and J. Walsh. 1998. "Industry and the Academy: Uneasy Partners in the Cause of Technological Advance." In R. Noll, ed., *Challenges to the Research University.* Washington, DC: Brookings Institution.

David, P. A., and D. Foray. 1995. "Accessing and Expanding the Science and Technology Knowledge Base." *STI Review* 16:13–68.

David, P. A., D. C. Mowery, and W. E. Steinmueller. 1992. "Analyzing the Economic Payoffs from Basic Research." *Economics of Innovation and New Technology* 2:73–90.

Eisenberg, R. 1996. "Public Research and Private Development: Patents and Technology Transfer in Government-Sponsored Research." *Virginia Law Review* 82:1663–727.

Evenson, R. E. 1982. "Agriculture." In R. R. Nelson, ed., *Government and Technical Progress: A Cross-Industry Analysis.* New York: Pergamon, 1982.

Griliches, Z. 1990. "Patent Statistics as Economic Indicators: A Survey." *Journal of Economic Literature.*

Hall, B. H., and R. H. Ziedonis. 2000. "The Patent Paradox Revisited: Determinants of Patenting in the U.S. Semiconductor Industry, 1979–1995." Forthcoming, *RAND Journal of Economics.*

Heilbron, J., and R. Seidel. 1989. *Lawrence and His Laboratory: A History of the Lawrence Berkeley Laboratory.* Berkeley, CA: University of California Press.

Henderson, R., A. B. Jaffe, and M. Trajtenberg. 1994, March 18–20. "Numbers Up, Quality Down? Trends in University Patenting, 1965–1992." Presented at the CEPR conference on "University Goals, Institutional Mechanisms, and the 'Industrial Transferability' of Research." Stanford University, Palo Alto, CA.

Henderson, R., A. B. Jaffe, and M. Trajtenberg. 1998. "Universities as a Source of Commercial Technology: A Detailed Analysis of University Patenting, 1965–88." *Review of Economics & Statistics* 80: 119–27.

Katz, M. L., and J. A. Ordover. 1990. "R&D Competition and Cooperation." *Brookings Papers on Economic Activity: Microeconomics 1990:* 137–92.

Kortum, S., and J. Lerner. 1997. "Stronger Protection or Technological Revolution: Which Is Behind the Recent Surge in Patenting?" Boston University. Unpublished manuscript.

Levin, R. C., A. Klevorick, R. R. Nelson, and S. Winter. 1987. "Appropriating the Returns from Industrial Research and Development." *Brookings Papers on Economic Activity:* 783–820.

Mowery, D. C. 1983. "The Relationship Between Intrafirm and Contractual Forms of Industrial Research in American Manufacturing, 1900–1940." *Explorations in Economic History:* 351–74.

Mowery, D. C., and N. Rosenberg. 1993. "The U.S. National Innovation System." In R. R. Nelson, ed., *National Innovation Systems: A Comparative Analysis.* New York: Oxford University Press.

Mowery, D. C., and N. Rosenberg. 1998. *Paths of Innovation: Technological Change in 20th-Century America.* New York: Cambridge University Press.

Mowery, D. C., and A. A. Ziedonis. 1998. "Market Failure or Market Magic? Structural Change in the U.S. National Innovation System." *STI Review* 22:101–36.

Mowery, D. C., and A. A. Ziedonis. "Academic Patent Quality and Quantity before and after the Bayh-Dole Act in the United States." Forthcoming, *Research Policy.*

Mowery, D. C., R. R. Nelson, B. Sampat, and A. A. Ziedonis. 1999. "The Effects of the Bayh-Dole Act on U.S. University Research and Technology Transfer: An Analysis of Data from Columbia University, the University of California, and Stanford University." In L. Branscomb and R. Florida, eds., *Industrializing Knowledge.* Cambridge, MA: MIT Press.

Mowery, D. C., R. R. Nelson, B. N. Sampat, and A. A. Ziedonis 2001. "The Growth of Patenting and Licensing by U.S. Universities: An Assessment of the Effects of the Bayh-Dole Act of 1980," *Research Policy* 30:99–119.

Mowery, D. C., and B. Sampat. 1999, March 30. "Patenting and Licensing of University Inventions: Lessons from the History of the Research Corporation." Presented at the conference on "R&D Investment and Economic Growth in the 20th Century." Berkeley, CA: U.C. Berkeley.

National Research Council. 1997. *Intellectual Property Rights and Research Tools in Molecular Biology.* Washington, DC: National Academy Press.

National Science Board. 1996. *Science and Engineering Indicators: 1996.* Washington, DC: U.S. Government Printing Office.

National Science Foundation. 1994. *National Patterns of R&D Resources: 1994.* Washington, DC: U.S. Government Printing Office.

National Science Foundation. 1996. *National Patterns of R&D Resources: 1996.* Washington, DC: U.S. Government Printing Office.

National Science Foundation. 1998. *National Patterns of R&D Resources: 1998.* Washington, DC: U.S. Government Printing Office.

National Science Foundation. 2000. *National Patterns of R&D Resources: 1999 Data Update.* Available at www.nsf.gov.80/sbe/srs/nsf00306/start.htm.

Nelson, R. R. 1959. "The Simple Economics of Basic Scientific Research." *Journal of Political Economy:* 67(3):297–306.

Nelson, R. R. 1992. "What Is 'Commercial' and What Is 'Public' About Technology, and What Should Be?" In N. Rosenberg, R. Landau, and D. Mowery, eds., *Technology and the Wealth of Nations.* Palo Alto, CA: Stanford University Press.

Nelson, R. R., and R. Mazzoleni. 1997. "Economic Theories about the Costs and Benefits of Patents." In National Research Council, *Intellectual Property Rights and Research Tools in Molecular Biology.* Washington, DC: National Academy Press.

Nelson, R. R., and P. Romer. 1997. "Science, Economic Growth, and Public Policy." In B. Smith and C. Barfield, eds., *Technology, R&D, and the Economy.* Washington, DC: Brookings Institution.

Office of Technology Transfer, University of California. 1997. *Annual Report: University of California Technology Transfer Program.* Oakland, CA: University of California.

Rosenberg, N. 1992. "Scientific Instrumentation and University Research." *Research Policy* 21:381–90.

Rosenberg, N. 1998. "Technological Change in Chemicals: The Role of University-Industry Relations." In A. Arora, R. Landau, and N. Rosenberg, eds., *Chemicals and Long-Term Economic Growth.* New York: John Wiley.

Rosenberg, N., and R. R. Nelson. 1994. "American Universities and Technical Advance in Industry." *Research Policy* 23:323–48.

Science. 1997, April 25. "Publishing Sensitive Data: Who's Calling the Shots?": 523–6.

Stata Corp. 1997. *Stata Statistical Software: Release 5.0.* College Station, TX: Stata Corporation.

Trajtenberg, M., R. Henderson, and A. Jaffe. 1997. "University Versus Corporate Patents: A Window on the Basicness of Inventions." *Economics of Innovation and New Technology* 5:19–50.

Williamson, O. E. October 1979. "Transaction Cost Economics: The Governance of Contractual Relations." *Journal of Law and Economics:* 22: 233–62.

Williamson, O. E. 1985. *The Economic Institutions of Capitalism.* New York: Free Press.

U.S. General Accounting Office. 1995. *University Research: Effects of Indirect Cost Revisions and Options for Future Changes.* Washington, DC: U.S. Government Printing Office.

U.S. Office of Management and Budget, Executive Office of the President. 1995. *The Budget of the United States Government for Fiscal 1996.* Washington, DC: U.S. Government Printing Office.

7

Should the Government Subsidize Supply or Demand in the Market for Scientists and Engineers?

Paul M. Romer, *Graduate School of Business, Stanford University*

Executive Summary

This paper suggests that innovation policy in the United States has erred by subsidizing the private sector demand for scientists and engineers without asking whether the educational system provides the supply response necessary for these subsidies to work. It suggests that the existing institutional arrangements in higher education limit this supply response. To illustrate the path not taken, the paper considers specific programs that could increase the numbers of scientists and engineers available to the private sector.

Preface

My son attends an undergraduate institution that specializes in science and engineering. A degree from this school will cost more than $100,000 in tuition and 4 years of his time. For parents and students who contemplate an investment of this magnitude, the school provides information about labor market outcomes for its graduates. On its web site, the school provides the median salaries for students who accepted jobs after graduation and the Ph.D. completion rates for students who go on to graduate school. If you search, you can see the entire distribution of outcomes—a listing for each student of the starting salary or graduate school in which they enrolled.

If my son pursues a doctoral degree after he graduates, he will have to make an even larger investment. Net of the various sources of support that are available to graduate students, the direct tuition cost of a doctorate will probably be less than the cost of his undergraduate degree. He will, however, have to invest another 4 to 8 years of foregone earnings, which will be substantially higher once he completes his undergraduate degree.

His college is unusual. Most undergraduate institutions do not provide any useful information about labor market outcomes for degree recipients. Yet as this example shows, it is perfectly feasible for a school to provide this kind of information. Given the stakes, it is even more surprising that many graduate programs in science and engineering also fail to provide this kind of information to prospective students. The paucity of information is obvious to anyone who is familiar with the graduate school application process, but to demonstrate it more formally, I asked a research assistant to begin the application process for the top 10 graduate departments of mathematics, physics, chemistry, biology, computer science, and electrical engineering in the United States. (The rankings were taken from US News and World Report.) For comparison, I also asked him to begin the application process to the top 10 business and law schools.

In response to his 60 initial requests for information from the science and engineering programs, he received not one response giving information about the distribution of salaries for graduates, either in the initial information packet or in response to a follow-up inquiry from him. In contrast, he received salary information for 7 of the 10 business schools in the application packet, and in response to his second request, he was directed to a web page with salary information by one of the three nonrespondents from the first round. (It is possible that the information could have been found on the web page for the other two business schools, but to maintain consistency in the treatment of the different programs, he did not look for more information if a school did not respond with directions about where to get it.) Four out of the 10 law schools gave salary information in the application packet and three more of them directed him to this information in response to a second request.

I.   Introduction

The most important economic policy question facing the advanced countries of the world is how to increase the trend rate of growth of output per capita. In the middle of the 20th century one might have argued that preventing depressions was the more urgent challenge, but at least in the advanced countries of the world, progress in macroeconomic stabilization policy has reduced the threat of a paralyzing economic collapse and even reduced the frequency of mild recessions. In this environment, the lure of better growth policy is compelling. If an economy can increase its trend rate of growth by even a small amount, the cumulative effect on standards of living is too big to ignore.

Many scholars and policymakers are convinced that during the 20th century, rapid technological progress in the United States drove the unprecedented growth in output and standards of living we enjoyed. In addition, they believe that this rapid rate of technological change was fostered by a publicly supported system of education that provided the essential input into the process of discovery and innovation—a steady flow of people trained in the scientific method and in the state of the art in their area of specialization.

If this interpretation of our recent past is correct, it follows that any proposal for achieving an even higher trend rate of growth in the United States should take full account of the detailed structure of our current system of higher education for natural scientists and engineers. Policymakers must recognize that these institutions exhibit puzzling features such as those described in the preface—an almost total lack of information on future market opportunities for students who enter their programs.

Unfortunately, in the last 20 years, innovation policy in the United States has almost entirely ignored the structure of our institutions of higher education. As a result, government programs that were intended to speed up the rate of technological progress may in fact have had little positive effect. We have undertaken major spending initiatives in the area of innovation policy, our most important area of economic policy, without subjecting their economic assumptions to even a cursory check for logical coherence or factual accuracy.

In what follows, I will point to the fundamental conceptual flaw behind the government programs that have been used in the last 20 years to encourage innovation in the private sector. These programs try to stimulate the demand for scientists and engineers in the private sector. To succeed, they depend on a positive supply response that the educational system seems incapable of providing. I will also describe a class of alternative policy programs that could be used to fill the gap created by an exclusive reliance on demand-side subsidies. These alternative programs return to an early style of government policy, one that works directly to increase the supply of scientific and engineering talent.

Section II below starts with a quick recapitulation of the reasons why decisions concerning innovation policy are so important for the economic well-being of future generations. Section III shows how a demand-side innovation policy such as a tax credit for research and development or a program of research grants affects the market for scientists and engineers. It shows why even a well-designed and extremely generous program of this kind will fail to induce more innovation and faster growth if the educational system does increase supply in response to changes in wages. Section IV provides an overview of trends in the supply of scientists and engineers. Sections V and VI look at the market for undergraduates and for Ph.D. recipients respectively. Section VII summarizes and interprets the evidence from these sections.

One of the surprising features of the political debate surrounding demand-subsidy policies is its narrow focus. Few participants in this debate seem to have considered the broad range of alternative programs. Section VIII tries to broaden the debate by suggesting feasible policies that could be considered. More specifically, it outlines a general process that policymakers could adopt for formulating growth policy. This process starts by distinguishing between goals and programs. To be specific, this section outlines four general goals that could guide the formulation of growth policy. The first possible goal that policymakers might adopt would be to target a specific increase in the number of students who receive undergraduate degrees in the natural sciences and engineering. A second would be to encourage more innovation in the graduate training programs that our universities offer to students who are interested in careers in science and engineering. A third would be to preserve the strength of our existing system of Ph.D. education. A fourth would balance amounts that the federal government spends on subsidies for supply and demand of scientists and engineers.

If policymakers in an economy were to adopt goals such as these, the next step would be to design specific programs that are intended to achieve these goals. In broadening the debate, this section also outlines three illustrative programs that policymakers could adopt to achieve these goals. The first is the introduction of training grants to universities that could be used to increase the fraction of undergraduates who receive degrees in natural science and engineering. The second is a system of exams that give objective measures of undergraduate achievement in natural science and engineering. The third is a new type of fellowship, backed by a substantial increase in funding, for students who continue their studies in graduate school.

The advantage of a process that separates goals from programs is that it establishes a natural way to evaluate any specific programs such as these. If the goals are precise and progress toward them can be quantified, then it should be easy to verify if any given program moves the economy closer to the goals. This makes it possible to experiment

GDP per Capita

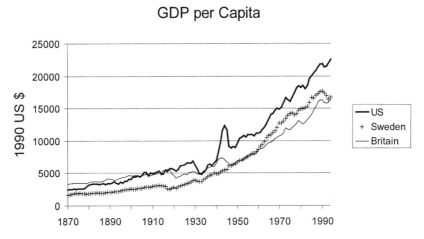

Figure 7.1
Income per capita from 1870 to 1992 for the U.S., Britain, and Sweden.
Data are from Angus Maddison (1995).

with a variety of programs, to expand the ones that work, and to shut
down the ones that do not.

II.   The Importance of Technology Policy

A quick look at the data in figure 7.1 suggests that there must be policy
choices, intentional or unintentional, that affect the trend rate of
growth. Using data assembled by Angus Maddison (1995), this figure
plots income per capita from 1870 to 1994 for the United States, Britain,
and Sweden. Over the century-and-a-quarter of data presented there,
income per capita grew in the United States at the rate of 1.8% per year.
In Britain, it grew at 1.3% per year. In the beginning of the sample period, the United States was a technological laggard, so part of its more
rapid growth could have come from a process of technological
catch-up with Britain, which was at that time the worldwide technology leader. But even at the beginning of this period, it was clear that the
United States was also capable of generating independent technological advances—for example, in the area of manufacturing based on the
assembly of interchangeable parts. (See Rosenberg 1969 for an account
of the reaction that the "American system of manufactures" caused in
Britain by the middle of the 19th century.) Moreover, as the United
States surged ahead of Britain in the 20th century, it maintained the

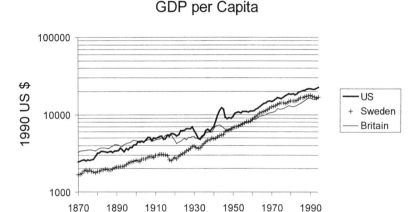

Figure 7.2
Same data as figure 7.1, on a logarithmic scale.
Data are from Angus Maddison (1995).

faster rate of growth that was apparent from the beginning. This is most clearly evident in figure 7.2. Because the data are plotted on a logarithmic or ratio scale, straight lines in the figure correspond to constant rates of growth. In the second half of the century, the rate of growth in Britain accelerated moderately. The rate of growth that had been initiated in the U.S. remained essentially unchanged. The policies and institutions in the United States made possible a trend rate of growth of income per person that was significantly faster than the trend that had pushed Britain into the position of worldwide leadership in the 19th century. Given the limited state of our knowledge of the process of technological change, we have no way to estimate what the upper bound on the feasible rate of growth for an economy might be. If economists had tried to make a judgment at the end of the 19th century, they would have been correct to argue that there was no historical precedent that could justify the possibility of an increase in the trend rate of growth of income per capita to 1.8% per year. Yet this increase is what we achieved in the 20th century.

The experience in Sweden suggests that even higher sustained rates of growth of income per capita can sometimes be possible. During the 50 years from 1920 to 1970, income per capita in Sweden grew at the much higher rate of 3% per year. Once again, this faster rate of growth could be due, at least in part, to the process of technological catch-up.

Moreover, growth in Sweden has slowed down considerably since 1970. Nevertheless, the experience in Sweden should at least force us to consider the possibility that if we arranged our institutions optimally, growth in the United States could take place at an even higher rate than that to which we have become accustomed. In the coming century, perhaps it will be possible to increase the rate of growth of income per capita by an additional 0.5% per year, to 2.3% per year.

The implications of a change of this magnitude would be staggering. For example, according to recent CBO estimates that were based on continuation of the historical trend rate of growth, in the year 2050 the three primary government entitlement programs—Social Security, Medicare, and Medicaid—will require an increase in government spending equal to about 9% of projected GDP. If the rate of growth over the next 50 years were to increase by 0.5% per year, GDP in 2050 would be 28 percentage points larger. By itself, faster growth could resolve all of the budget difficulties associated with the aging of the baby boom generation, and still leave ample resources for dealing with any number of other pressing social problems. And of course, the longer a higher growth rate can be sustained, the larger the effect it will have. By the year 2100, the additional 0.5% per year would translate into a GDP that is 1.65 times as large as it would otherwise have been.

Other types of evidence suggest that an increase in the rate of growth of 0.5% per year is not beyond the realm of possibility. In his survey of returns to investment in R&D, Zvi Griliches (1992) reports a wide range of estimates for the social return, with values that cluster in the range of 20% to 60%. Take 25% as a conservative estimate of the social return on additional investment in R&D. If we were to increase spending on R&D by 2% of GDP (and maintain the same rate of return on our investments in R&D—more on this in the next section) the rate of growth of output would increase by the hoped-for 0.5% per year. If the true social return is higher, say 50%, the extra investment in R&D needed to achieve this result would be correspondingly lower, just one additional percent of GDP. These estimates are also consistent with other estimates, which suggest that the level of resources currently devoted to research and development may be far below the efficient level. For example, after they calibrate a formal growth model to the results from micro level studies of the productivity of research and development, Chad Jones and John Williams (1998) calculate that the optimal quantity of resources to devote to research and development could be four times greater than the current level.

There is another way to look at estimates of this kind, one that is closer to the spirit of the analysis that follows below. GDP in the United States is about $10 trillion dollars. One percent of this would be $100 billion per year in additional spending on R&D. If it costs $200,000 per year to hire and equip the average worker in this sector, this means that we would need to increase the stock of workers employed in R&D by roughly 500,000. The question that policymakers must confront if they are serious about increasing the amount of R&D that is performed is where these additional high-skilled workers will come from.

There is no certainty that growth would necessarily speed up even if we did undertake all the right steps in an effort to do so. There is ambiguity in the historical record, and even if there were not, there is no guarantee that relationships that held in the past will continue to hold in the future. Moreover, even in the best case, we should recognize that there might be substantial lags between the initiation of better policy and the realization of faster output growth. For example, one highly successful example of a government policy that did increase the rate of technological change was the creation of the new academic discipline of computer science in the 1960s. (See Langlois and Mowery 1996 for a discussion of the episode.) Even now, with the passage of 40 years, our sense of the magnitude of the payoff from this investment is still growing.

Notwithstanding all these caveats, a possibility, even a remote possibility, of a change as profound as another permanent 0.5% increase in the trend rate of growth in the world's leading economy ought to excite the imagination. Compared to this, even landing a man on the moon would seem a minor achievement. One would think that this kind of possibility would inspire us to try new things, to make every effort to understand what will work and what will not as we strive for this goal. By this kind of standard, the efforts we have made in the last 2 decades have been remarkably timid and poorly conceived.

III. Demand Subsidies

Unless one is careful and makes use of some simple economic theory, it is easy to fall into an all-too-common trap in discussions about innovation policy. The key point was signaled above in the switch from a discussion of spending on R&D to a discussion of the number of workers engaged in this activity. To speed up growth, it is not enough to increase *spending* on research and development. Instead, an economy must increase the total *quantity of inputs* that go into the process of research and development. Spending is the product of a quantity and a price. To simplify the discussion, assume for now that people—scientists and engineers—are the key inputs in research and development. Formally, let E stand for spending on research and development and let N represent the number of scientists and engineers working in this area. Let w represent the average wage for these workers. Then trivially, $E = N \times w$. An increase in expenditure E will not necessarily translate into a corresponding increase in N, the number of scientists and engineers engaged in R&D. In principle, it is entirely possible for the entire increase in E to pass through as increases in w.

Continuing with the simplifying assumption that scientists and engineers are the only inputs in the production of research and development, we can illustrate how w is determined using the simple supply and demand framework presented in figure 7.3. The horizontal axis measures the number of scientists and engineers working in the private sector on R&D. The vertical axis measures their wage. The downward-sloping demand curve indicated by the solid line represents the

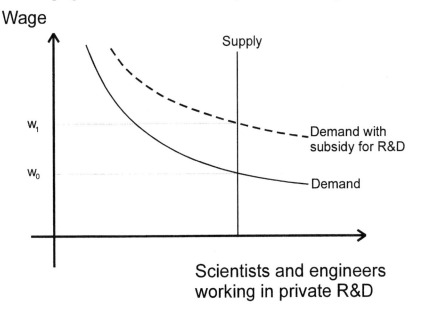

Figure 7.3
Wage w versus number of scientists and engineers working in private R&D

private return captured by a firm that hires some additional scientific workers and undertakes more research and development.

In the figure, the supply of scientists and engineers is represented by a vertical supply curve. The vertical slope of the curve reflects an assumption that the number of young people who become scientists and engineers and go to work in the private sector does not adjust in response to an increase in the wage that they receive for employment in the private sector. Section IV below outlines a much more complicated picture of the supply response of our educational system, but it is useful to start with the simple case of zero supply elasticity. To motivate this assumption, it is enough to keep the story from the preface in mind. The lack of information that is available to students who are making decisions about careers in science and technology suggests that our existing educational institutions may not lead to the kind of equilibration that we take for granted in many other contexts. If students do not have information about what wages will be, it will be much harder for them to adjust their career decisions in response to wage changes.

The downward-sloping dashed line in the figure represents the private demand for research workers when the government provides a subsidy for R&D. This subsidy could take the form of special tax advantages such as those afforded by the research and experimentation tax credit offered in the United States. Alternatively, the subsidy could take the form of cash payments to some firms as part of a cost-sharing agreement in which the government pays part of the cost of a research and development program. This is the kind of subsidy offered by partnership programs such as the Small Business Innovative Research (SBIR) grant program administered by the Small Business Administration or the Advanced Technology Program (ATP) administered by the Department of Commerce. Whether it comes in the form of tax credits or research grants, the effect of government spending is to shift up and to the right the demand for scientists and engineers who can perform the R&D.

From the perspective of a single firm, it seems obvious that a special tax incentive or a research grant will encourage the firm to hire more scientists and engineers and thereby to cause more inputs to be devoted to R&D. Yet one of the most basic insights in economics is that for the economy as a whole, things have to add up. If the total number of scientists and engineers is fixed, it is arithmetically impossible for employment of scientists and engineers to increase at all firms. As illustrated in the figure, if the supply curve of scientists and engineers is fixed, then the increase in demand induced by the subsidy will translate into a proportional increase in wages for scientists and engineers with no increase in the inputs that are devoted to R&D.

It is important to recognize that this argument is separate from the usual concerns about "additionality" that have been raised with respect to R&D demand subsidy programs. People who focus on this problem are worried about how much the demand curve shifts. That is, they are worried that an additional dollar in subsidies does not translate into much additional private spending on R&D. This is a nontrivial issue. The evidence does seem to suggest that more generous tax treatment for R&D leads to higher *reported* levels of spending on R&D at firms. (See for example Bronwyn Hall and John van Reenen 1999.) An additional dollar in tax benefits seems to lead to about one additional dollar in reported R&D expenditure by firms. However, there is much less evidence about the extent to which this increase in reported R&D spending represents a true increase in spending relative to that which would have taken place in the absence of the credit. It is quite possible that some of this spending comes from relabeling of spending that would have taken place anyway. Deciding what qualifies for this credit is apparently a nontrivial problem for the tax authorities. Between 20% and 30% of claimed expenditures by firms are disallowed each year (National Science Board 1998:4-48).

For the SBIR program, Josh Lerner (1999) finds that firms that receive grants from the government experience more rapid sales and employment growth than a comparison group of firms selected to be similar to the recipient firms. This could be an indication that firms that receive grants do devote more inputs to R&D. But it could also reflect unobserved, intrinsic differences between the control group, which was constructed ex post by the researcher, and the recipient group, which was selected on the basis of a detailed application process that was designed to select particularly promising firms. In related work, Scott Wallsten (2000) finds that firms that receive a research grant from the government under the SBIR program seem to substitute these grant funds for other sources of funds, with little or no net increase in spending on R&D.

For both the tax credit and direct grant programs, we can identify a coefficient m which measures the true increase in private spending on R&D associated with each additional subsidy dollar from the government. In each case, there is some uncertainty and debate about how large this coefficient is. But for any positive value of m, the argument outlined above shows that the entire increase in spending may show up as higher wages for the existing stock of workers, with no increase in the actual quantity of research and development that is performed. As a result, even a well-designed and carefully implemented subsidy could end up having no positive effect on the trend rate of growth for the nation as a whole.

Recent work by Austan Goolsbee (1998) suggests that, at least in the short run, the wage changes implied by a weak supply response are apparent in the data. He compares census data on wages for research workers with time series data that capture the variation in government spending on R&D. Direct government spending is well suited for this kind of analysis because it does not suffer from the concerns about additionality that are present for government subsidies for R&D. Surprisingly, using only these crude data, he finds strong effects on wages. For example, during the defense build-up between 1980 and 1984, federal spending on R&D increased, as a fraction of GDP, by 11%. His estimates suggest that this increased wages for physicists by 6.2% and aeronautical engineers by 5%.

In the face of this argument, defenders of demand-side R&D subsidies can respond in three ways. First, they can argue that people are not the only inputs used in R&D. If other inputs such as computers and specialized types of laboratory equipment are supplied elastically, then government subsidies for R&D could increase the utilization of these other inputs even if the number of scientists and engineers remains constant. If this were truly the intent of the various subsidy programs, it would be much more cost-effective for the government to provide the subsidies directly for these other inputs. Salaries account for the majority of total R&D spending. For example, in university-based research, annual research expenditures on equipment during the last decade have varied between 5% and 7% of total research expenditures (National Science Board 1998). If the goal of the subsidy program were to increase the equipment intensity of research and development and if the ratio of spending on equipment in the private sector is comparable to the figure for universities, a special tax subsidy for the purchase of equipment used in research would be substantially less costly than one that is based on total expenditures including salaries. Similarly, the government could achieve substantial savings, and still increase the use of equipment in R&D, if it restricted the grants provided by the SBIR and ATP programs so that these funds could be used only for additional purchases of equipment.

In the case of the targeted grant programs administered by the ATP or the SBIR, a defender could argue that even if the existing research subsidies do not increase employment of scientists and engineers in the economy as a whole, they can increase employment at the recipient firms, at the cost of a reduction in employment at other firms. If government agencies were able to identify an allocation across firms and projects that is better than the one the market would implement, the targeted grant programs could still be socially valuable. But even the strongest supporters of the subsidy programs are hesitant to make this kind of claim about the superiority of government allocation processes. Note also that because the research and experimentation tax credit is available to all firms, it cannot be justified on this kind of basis of any hypothesized ability of the government to improve the allocation of research inputs between firms and projects.

If the goal is not to encourage equipment investment in the R&D sector or to give the government a bigger role in deciding how to allocate scarce R&D personnel, some other motivation must lie behind these spending programs. The final response could be for a defender of these programs to dispute the basic assumption behind the supply-and-demand model outlined here and argue that, at least in the long run, the supply of scientists and engineers working in R&D in the private sector does respond to demand-induced changes in wage. But to make this case, one must confront some of the peculiar features of the educational system that actually produces these highly skilled workers and ask if there are more cost-effective ways to increase the supply of these types of workers.

IV.   Overview of the Supply of Scientists and Engineers

Figures 7.4 and 7.5 give a broad overview of trends in the supply of scientists and engineers in the United States. Figure 7.4 updates data presented by Chad Jones (1995) on the number of scientists and engineers in the United States who are employed in research and development. These data are scaled by the size of the labor force. They show an increase in R&D employment as a fraction of the labor force from about 0.3% of the labor force in 1950 up to about 0.8% in the late 1960s, with no strong trend thereafter. The underlying data for this figure are collected by the NSF. (Data since 1988 are taken from Table 3-15 from National Science Board 1998.)

Official statistics on formal research and development capture only part of the private sector effort directed at innovation. Also, no consistent data series on employment in R&D is available in years prior to 1950. To give a more comprehensive overview of the proportions of skilled workers in the labor force over a longer time horizon, figure 7.5

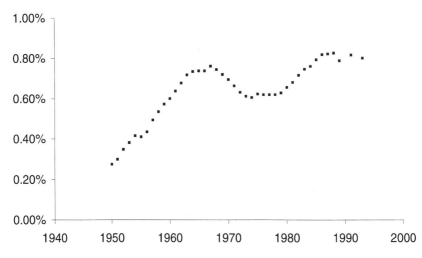

Figure 7.4
Scientists and engineers in R&D as a fraction of the labor force
Data from Chad Jones 1995.

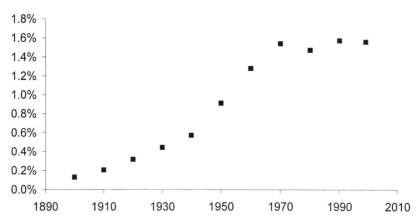

Figure 7.5
Engineers as a fraction of the labor force
Data are from Historical Statistics of the United States, 1975, and Statistical Abstract of
the United States, various years.

presents data on the total number of engineers as a fraction of the labor force, using occupational data collected by the Bureau of the Census. This series shows a similar pattern. Engineers increase, as a fraction of all workers, from the turn of the century up until 1970, and remain roughly constant thereafter.

Taken together, these figures offer little reassurance that the aggregate supply of scientists and engineers responds efficiently to market demand. Of course, it is logically possible that the growth in the demand for scientists and engineers experienced a sharp fall starting in the late 1960s. However, other labor market evidence based on relative wages such as that presented by Katz and Murphy (1992) suggests that a process of skill-biased technological change that raised wages for skilled relative to unskilled workers continued at about the same pace in the 1970s and 1980s as in the 1960s. Other work (see for example, Autor, Katz, and Krueger 1998) suggests that, if anything, the rate of skill-biased technological change actually increased in the period from 1970 to 1995 relative to the period from 1940 to 1970. Taken together, these data on quantities plus the independent evidence on the demand for skill suggest that one look more carefully at other possible factors that could influence the supply of scientists and engineers.

Figure 7.6 gives a schematic outline of the process that actually determines the supply of scientists and engineers. The two key stages in the production process are undergraduate education and graduate education. (For simplicity, graduate programs that lead to a terminal master's degree are grouped in this figure with those that provide Ph.D. level training.) The first major branch in the process distinguishes undergraduates who receive degrees in the natural sciences or engineering (NSE degrees) from those who receive all other types of degrees. Section V below looks at the possible nonmarket forces that could constrain this decision. After a student receives an undergraduate NSE degree, she can either go to work in the private sector or continue on to receive graduate training. Section VI looks at recent developments in the market for people with an advanced degree in the natural sciences or engineering.

V. The Supply of Undergraduate Degrees in Science and Engineering

The market for education suffers from pervasive problems of incomplete information. Students contemplating a choice between different institutions typically have very little information about the

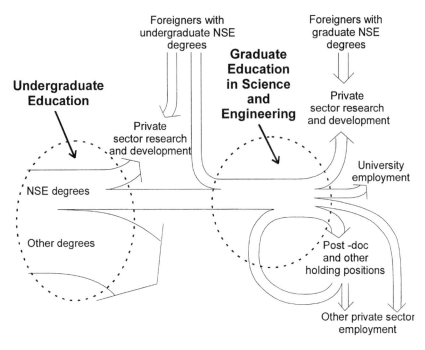

Figure 7.6
Schematic outline of the process determining the supply of scientists and engineers.

value-added they can expect to receive from one institution versus another. Employers selecting among graduates from different institutions also have very little objective basis for judging the absolute achievement levels of students from different schools, or even about students from the same school who have followed different courses of study. The competitive strategy that seems to have emerged in this market is one where undergraduate institutions have developed extensive systems for screening students by ability level. They enroll the most able students that they can attract. The schools compete for these students in large part by publicizing the degree of selectivity. Students, in turn, compete for admission to the most selective institutions because a degree from a more selective institution offers a stronger signal about the student's ability. Using data from the different campuses of the University of California, Robert Frank and Philip Cook (1995) suggest that competition along these lines has been getting more intense. For example, over time, SAT scores for students attending Berkeley, the UC campus that is perceived to be the most selective, have been increasing relative to SAT scores at other campuses.

In this kind of competitive environment, the traditional liberal arts university may face little pressure to respond to changing market demands for different types of skills. For example, imagine that government subsidies increased the market wage for scientists with several years of training beyond the undergraduate degree. Imagine that students are somehow informed about this change in the wage and respond by increasingly enrolling in undergraduate science courses that will prepare them for further study in engineering and science. A liberal arts university that has a fixed investment in faculty who teach in areas outside of the sciences and that faces internal political pressures to maintain the relative sizes of different departments may respond to this pressure by making it more difficult for students to complete a degree in science. Faculty in the departments that teach the basic science courses will be happy to "keep professional standards high" and thereby keep teaching loads down. Faculty in other departments will be happy to make study in their departments more attractive, for example by inflating the average grade given in their courses.

There is clear evidence that this kind of response currently operates on campuses in the United States. First, the number of students who begin their undergraduate careers with the intent of receiving a degree in science and engineering is substantially higher than the number who actually receive such a degree. For example, for white students, 12% of entering students intend to major in natural sciences and 9% plan to major in engineering. Only 8% of graduating students actually receive a degree in natural sciences and only 5% receive a degree in engineering (National Science Board 1998). For minority students, the attrition rate is even higher.

One additional indication of the pressure to shift students out of science and engineering degrees comes from the difference in the distributions of grades offered in courses required for degrees in these areas as opposed to grades in other courses of study. Measuring this difference is not straightforward because even within a department such as mathematics, and even within a specific subject area such as linear algebra, there are courses with easier grade distributions that are intended for people who will not continue toward a degree in science, and courses with a lower distribution of grades for people who will.

For example, students who place out of the basic calculus course on the basis of an advanced placement exam are more likely to take more difficult math courses than students who do not. This tends to lower the average grade they receive in the second-level math courses that they take. If one does not correct for this fact, one finds that math grades for the students who place out of calculus are not, on average, any higher than math grades for students who do not place out of calculus. However, if one holds constant the specific second-level math courses that students take and compares grades for students with different backgrounds who take the same course, it is clear that students who have placed out of calculus do receive higher math grades than other students taking the same class (Rick Morgan and Len Ramist 1998).

To do this kind of analysis, the College Board, which administers the advanced placement exam, collected data from a representative sample of 21 selective universities. Using these data, one can do a direct comparison of grade distributions across different fields of study. Take, for example, the sample of second-level math courses that students who place out of calculus attend. These tend to be biased toward the classes that students majoring in mathematics or the natural sciences will take. One can then compare the distribution of grades in these courses with the distribution of grades in second-level English courses taken by students who receive advanced placement credit in English composition; or with the distribution of grades in second-level history courses taken by students who receive advanced placement credit in American history. As table 7.1 shows, in the selected math courses, 54% of all students received a grade of A or B. For the English courses, the fraction with an A or a B is 85%. For the history courses, the fraction is 80%. For social science courses such as political science or economics, the fraction of students who receive a grade of either A or B is about 75%.

As figure 7.6 shows, immigration is an alternative source of supply for the labor market in the United States. If the domestic supply of scientists and engineers is constrained to a significant extent by our existing system of undergraduate education, one should see evidence that the response in terms of undergraduate NSE degrees differs from that of immigrants. Recently, much of the discussion of migration has focused on political pressure from technology-intensive firms for increases in the number of H1B visas that permit private firms to hire skilled workers from abroad to fill entry-level jobs in areas such as computer programming. This debate has obscured the extremely important role that immigration has long played in supplying scientists and engineers with the highest levels of skill. Moreover, immigration is clearly responsive to demand conditions. Fields such as computer science and engineering, where indicators suggest that market demand is high relative to the available supply, are the ones that have experienced

Table 7.1
Fractions of students receiving an A or a B in different subjects

Subject Area	Fraction of Students Receiving a Grade of A or B
English	85
History	80
Economics and Political Science	75
Mathematics	54

the largest inflows from abroad. For example, in 1993, 40% of the people in the United States who had a Ph.D. degree in engineering were foreign-born. In computer science, 39% of the Ph.D. holders were foreign-born. In the social sciences, where demand for new Ph.D. recipients is generally much lower (economics being a notable exception), only 13% of the Ph.D. holders were foreign-born (National Science Board 1998). These immigration flows stand in sharp contrast to the trends in undergraduate education. From the mid-1980s until 1995, the number of undergraduate degrees in engineering and in mathematics and computer sciences fell substantially. For example, in the 1980s and 1990s, as the personal computer and internet revolutions were unfolding, the number of undergraduate degrees in computer science showed no strong trend, increasing at first in the mid 1980s, then falling in the 1990s and ending at about the level at which it started in the early 1980s.

Engineering degrees follow a very similar pattern (National Science Board 1998: Table 2-20). Between 1981 and 1995 there is no change in the number of undergraduates who receive degrees in engineering. The number does increase in the late 1980s but then returns to the previous level. For future reference, note that the number of master's degrees in engineering behaves quite differently. From 1981 to 1995 the total number of master's degrees in engineering increased steadily so that the number in 1995 was about 1.7 times the number in 1981 (National Science Board 1998: Table 2-27).

Another sign that the domestic enrollment of students who are able to continue in science and engineering is a critical bottleneck comes from an examination of downstream developments in Ph.D. education. From the mid-1970s to the mid-1980s, the number of Ph.D. degrees awarded in the United States each year in natural sciences and engineering remained roughly constant at about 12,500. (Here, as elsewhere, natural sciences and engineering exclude behavioral and social sciences.) Then, starting in 1986, this number began a steady increase up to 19,000 per year in 1995.

We can use this expansion in the size of Ph.D. programs to gauge the elasticity of the foreign supply response and compare it to the domestic supply response. In 1986, U.S. citizens accounted for about 8,000 of Ph.D. degree recipients, and noncitizens accounted for the other 4,500. In 1995, the number of degrees for U.S. citizens had increased by about 20% to around 10,000 and the number of degrees awarded to noncitizens had more than doubled to 9,000 (National Science Board 1998: Table 2-35).

A similar, though less extreme picture emerges from an examination of master's degrees, particularly in the high-demand areas of computer science and engineering. As market opportunities for holders of the master's degree increased and universities added to the number of slots that they made available in master's degree programs, foreign students responded more strongly than U.S. citizens, just as they did when new positions in Ph.D. programs opened up. In 1975, foreign students received 22% of the master's degrees in engineering and 11% of the master's degrees in math and computer science. By 1995, foreign students accounted for 39% of the master's degrees in engineering and 35% of the master's degrees in math and computer science (National Science Board 1998). In both instances, increased downstream demand for undergraduates with NSE degrees does not seem to have induced a sufficient supply response. The system equilibrated by importing more foreigners.

VI.   The Supply of Ph.D. Degrees in Science and Engineering

The sharp increase in the 1990s in the number of Ph.D.s granted has been accompanied by generally declining job prospects for degree recipients. In the most recent period, it is possible that part of the reason why undergraduate students did not pursue degrees in the natural sciences is that they were vaguely aware of the worsening job prospects that Ph.D. recipients faced. Note, however, that developments in the academic market for Ph.D.s cannot explain the absence of an increase in undergraduate degrees in engineering or in specialized areas such as computer science where job prospects for Ph.D. recipients have remained strong. Also, the weak market for new Ph.D.s would only have been a factor fairly recently, primarily since 1990 when the increased supply of Ph.D. recipients began to show up on the market. Nevertheless, going forward, the weak academic market for some types of Ph.D.s will certainly be a complicating factor in any attempt at increasing the number of undergraduate degrees that are awarded in natural science and engineering. To increase the number of undergraduates who receive an undergraduate degree in the natural sciences and engineering, they must be convinced that this kind of degree can lead to better career outcomes than the dead-end postdoctoral positions that have become increasingly common in some fields.

Independent of its role in influencing undergraduate degrees in the United States, understanding the behavior of the market for Ph.D.s is critical to the formulation of policy concerning the supply of scientists and engineers. The thrust of the possible programs outlined below is to substantially increase this supply. Yet many people in the academic community are convinced that the most pressing science policy issue in the United States is the Ph.D. glut. They have advocated measures that would reduce the supply of Ph.D.-level scientists and engineers. A careful look at the market for Ph.D.s is necessary to explain why increases in the supply of scientists and engineers with several years of graduate training are still called for even in the face of difficulties in the labor market for Ph.D.s. The key point here is to distinguish between people who are trained exclusively for employment in research universities and people who can work in research and development in the private sector.

Look again at figure 7.6. Events in the Ph.D. market can be summarized in terms of this figure. As noted above, the total flow of students through NSE Ph.D. programs increased starting in the late 1980s and continuing through the 1990s. Much of this flow has been directed at two of the alternatives upon leaving graduate school—university employment and postdoc and other holding positions. The challenge in this area is not to increase the total numbers of Ph.D. degree recipients, but to increase the fraction of them that can put their skills to work in private sector research and development.

This pattern of outcomes—increased numbers of Ph.D. recipients and steadily worsening academic job prospects—can be explained by increased subsidies for Ph.D. training. These subsidies derived from increased support for university-based research, which is complementary to Ph.D. training. As a result, the nature of the support for graduate students changed along with the level. Consider the sample of students who received their primary means of support for their Ph.D. education from the federal government. Between 1980 and 1995, the fraction whose primary mechanism of support was a traineeship fell from 25% to 15% and the fraction whose primary mechanism was a research assistantship increased from 55% to 63%. The fraction receiving their primary support from fellowships stayed roughly constant at about 10%. Among students whose primary support was from sources other than the federal government (primarily state governments), research assistantships also increased by about 10 percentage points (National Science Board 1998: Chapter 5).

Because this increase in supply consisted of people who planned to pursue academic research appointments, the increased supply of Ph.D. recipients was accompanied by generally worsening job prospects for Ph.D. recipients in the academic market. For example, consider in any year the sample of people with degrees in the natural sciences and engineering who were working in academic institutions and who had received their Ph.D. degree within the previous 3 years. In the early 1980s, there were about 17,000 of these recent degree recipients working in academic institutions. About half of them had faculty jobs. The rest held postdoctoral positions or some other form of appointment. By 1995, this same measure of recent Ph.D.s in academic institutions had increased to 23,000, but the number holding faculty positions remained roughly constant, at about 8,500. The entire increase of 6,000 recent degree recipients is accounted for by increases in nonfaculty appointments (National Science Board 1998: Table 5-29).

The problems in the academic market in the life sciences were documented in a report from the National Research Council (1998). In the last decade, this is the area that has benefited from the most rapid rate of growth of federal research support. Between 1970 and 1997, the median time to receipt of a degree increased by 2 years to a total of 8 years. The number of people who hold a postdoctoral appointment 3 or 4 years after receipt of the Ph.D. increased from 6% to 29% between 1973 and 1995. The fraction of Ph.D. recipients who do not hold a permanent full-time job in science and engineering 5 or 6 years after they have received their degree increased from 11% in 1973 to 39% in 1995. The 1995 data, which were the most recent available at the time that the National Research Council wrote its report, reflect long-term outcomes for the 1989–90 cohort of Ph.D. recipients. Because of the steady increase in the number of degree recipients throughout the 1990s, the competitive pressures in this field have probably worsened still further.

VII.   An Interpretation of the Evidence Concerning Higher Education

The picture that emerges from this evidence is one dominated by undergraduate institutions that are a critical bottleneck in the training of scientists and engineers, and by graduate schools that produce people trained only for employment in academic institutions as a side effect of the production of basic research results. This description of the system as a whole hides a heterogeneous mix of different types of institutions. Not all of them will behave according to the description given above.

For example, the pressure to keep enrollments down in the natural sciences and engineering will not be present at institutions that specialize in this kind of training. They may therefore face different kinds of incentives and behave differently in the competition for students. The institution that my son attends, Harvey Mudd College, is one of these specialized institutions, and this may explain why it features information about the market outcomes for its graduates more prominently than traditional liberal arts universities. A quick check of data from other schools is consistent with this observation. MIT and Caltech, two selective schools that also concentrate in science and engineering, present information about median salaries for their undergraduates on the web pages that provide information for potential applicants. Harvard and Stanford, two comparably selective institutions that cover the whole range of academic disciplines, apparently offer no information on their web pages about salaries or enrollments in graduate school.

One natural question that the model outlined here does not address is why competition by entry of more schools like Harvey Mudd, MIT, and Caltech has not partially solved the bottleneck problem described here. Mudd, which is about 50 years old, is a relatively recent entrant in this market, but in general, entry seems to be a relatively small factor in the competition between undergraduate institutions. Presumably the incomplete information available to students and employers about the quality of the education actually provided at any institution is a big factor limiting the entry process, but the nature of competition between schools deserves more careful consideration.

There are also different types of institutions that provide graduate education. The description offered here focuses primarily on graduate education in the sciences, which takes place almost exclusively within institutions where the revenue and prestige associated with research are more important motivating forces than tuition revenue. Training in these departments differs sharply from the kind of training offered by professional schools where income from tuition is a much more important determinant of institutional incentives. It should not be surprising that, as my research assistant discovered, business schools and law schools follow very different strategies from the ones used by departments of science when they compete for students. In many ways, master's level training in engineering is like these professional schools. Much of the income associated with these programs comes from tuition. Departments that get to keep a portion of this master's level, but not of undergraduate tuition revenue, should therefore be willing to expand the size of their master's programs at the same time that they put limits on the size of their undergraduate programs. These kinds of incentive effects may help explain why master's degree programs in engineering have shown steady growth while undergraduate engineering degrees have not.

In its report on career prospects in the life sciences, the National Academy Board on Biology concluded that policymakers should restrain the rate of growth of graduate students in the life sciences. In my language (not theirs) they also recommended that graduate education in the life sciences be reshaped along lines that are closer to those followed by professional schools. They recommended that students be given more information about career prospects, that they be given training that prepares them for employment in jobs outside of university-based research, and that funds that support the training of graduate students be shifted away from research assistantships and toward training grants or other forms of support that give more control over a student's education to the student.

This last and most controversial recommendation is the one that has the greatest potential to shift the traditional science-based model of graduate education closer to the model that we see in master's level professional schools of tuition-paying customers who collectively can exert a significant degree of control over what happens during the process of education. Similar proposals for modifying Ph.D. training have been made by a variety of study panels. All have received mixed support at best from the scientific community as a whole. (See the discussion of this point in National Science Board 1998: 5-33.)

Opposition to any change in the form of support for graduate education is usually justified in public on the grounds that there is insufficient evidence about what the effects might be for any change in the system of funding for graduate students. A more fundamental problem—one that goes largely unreported in print but that prominent scientists are willing to justify in private—is that the current funding and training system, one that puts graduate students in the position of apprentices to established scientists and that does not prepare students for careers outside of science, is crucial to the maintenance of the institutions of academic science. Recent work by Scott Stern (1999) offers convincing evidence that recipients of Ph.D. degrees exhibit a strong preference for engaging in the activities in science and are willing to accept substantial wage reductions if doing so will allow them to continue to pursue these activities. This preference could be the result of a selection process that attracts people with this taste into Ph.D. training in science, a training process that cultivates this taste, or a combination of the two. Regardless of the mechanism, any attempt to make the training of Ph.D. students resemble more closely the training of students in business schools could have the effect of significantly undermining the commitment to the ideas and process of science that Stern is able to document. This commitment, which may be psychologically and functionally similar to the commitment induced by training for membership in a religious order or a military unit, may be critical to the preservation of the institutions of science. Unfortunately, it may also help explain why the existing system of graduate education seems so poorly suited to training people for employment outside of academic science. For this combination of reasons, the task of modifying the educational system that trains scientists and engineers may be both very important and very delicate.

VIII.   Goals and Programs

To formulate growth policy, policymakers may want to start by distinguishing goals from programs. Goals should be conservative. They should represent objectives that are neither risky nor radical and for which there is a broad base of intellectual and political support. Goals should remain relatively constant over time. They should also imply metrics for measuring success. By these criteria, increasing the long-run trend rate of growth is not specific enough to be a goal. It is appropriately conservative and should be the subject of a broad consensus, but because it is so difficult to measure the trend rate of growth, it does not imply any workable metrics that we can use to measure progress toward the goal. In contrast, increasing the fraction of young people in the United States who receive undergraduate degrees in science and engineering could qualify as a goal. So could increasing the total quantity of resources that are devoted to research and development.

In contrast to a goal, a program is a specific policy proposal that seeks to move the economy toward a specific goal. For example, the Research and Experimentation tax credit is a specific program that is designed to achieve the goal of increasing the resources used in research and development. It should be possible to judge the success of a program against the metric implied by the goal that it serves. All programs should be designed so that they can be evaluated on a policy-relevant time horizon. If they are, they can also be less conservative and more experimental than the underlying goals. A variety of programs could be tried, including ones where there is some uncertainty about whether they will succeed. If the evidence shows that they do not work, they can be modified or stopped.

To illustrate how this framework could facilitate better analysis of the growth process, it helps to focus on a specific set of hypothetical goals. Imagine that policymakers and the public at large accepted the following goals because they want to increase the long-run rate of growth in the United States. (1) Increase the fraction of 24-year-old citizens of the United States who receive an undergraduate degree in the natural sciences and engineering from the current level of 5.4% up to 8% by the year 2010 and to 10% by 2020. (2) Encourage innovation in the graduate training programs in natural science and engineering. (3) Preserve the strengths of the existing institutions of science. (4) Redress the imbalance between federal government subsidies for the demand and supply of scientists and engineers available to work in the private sector.

Each of these goals suggests natural metrics for measuring progress. The NSF currently measures the fraction of 24-year-olds who receive undergraduate degrees in the natural sciences and engineering (NSE). These data are also available for other countries. Although the United States provides undergraduate degrees to a larger fraction of its young people than almost all other developed nations, many fewer undergraduates in the U.S. receive degrees in natural science and engineering. As a result, the fraction of all 24-year-olds with undergraduate NSE degrees is now higher in several nations than it is in the U.S. The United Kingdom (8.5%), South Korea (7.6%), Japan (6.4%), Taiwan (6.4%), and Germany (5.8%) all achieve levels higher than the 5.4%

level attained in the United States (National Science Board 1998: Chapter 2). The experience in the United Kingdom also shows that it is possible to expand this fraction relatively rapidly over time. In 1975, the figure there stood at only 2.9%.

The indicators for the next two goals will have to be more eclectic. Possible indicators of innovation in graduate education could include the creation of graduate training programs in new areas (bioinformatics, for example) where the private sector demand for graduates is high; or programs that involve new types of training (internships in private firms, perhaps); or programs that offer different types of degrees from the traditional master's or Ph.D. One would also like to see continued strength in the Ph.D. programs that form the core of our system of basic scientific research, measured perhaps by the quality of students that they attract both domestically and from abroad. The second and third goals explicitly allow for the possibility that developments in these two areas need not be closely linked. Universities might introduce new programs in an area such as bioinformatics that train people primarily for work in the private sector without affecting existing programs in biology. The new programs could have the same independence from Ph.D. training in biology that programs of chemical engineering have from Ph.D. training in chemistry. As a result, innovation in the sense of new programs need not imply any changes in the existing Ph.D. training programs and need not take any funding from those programs. If the country makes progress toward the first goal, and the number of U.S. citizens who pursue undergraduate studies in science increases, this could improve the quality of the domestic applicant pool for the traditional Ph.D. programs at the same time that it supplies people to the new alternative forms of graduate education.

It will take new funding from the federal government to encourage the introduction of new training programs and still preserve the strength of existing graduate programs. The last goal sets a rough benchmark that policymakers might use to set expectations for how much funding might be allocated on a permanent basis toward these goals. In the last 2 decades, the primary programs that have subsidized the private sector demand for R&D have been the research and experimentation tax credit, the SBIR program, and the ATP program. Rough estimates of the costs for these programs are $1 billion each per year for the tax credit and the SBIR program and between $300 and $400 million per year for the ATP program. The fourth goal suggests a starting target of around $2–2.5 billion per year in subsidies for the supply of scientists and engineers.

If policymakers adopted these kinds of goals, then it would be a straightforward process to design programs that might help achieve them and to evaluate these programs after they are implemented. The following list of programs only begins to suggest the range of possibilities that could be considered. (1) Provide training grants to undergraduate institutions that are designed to increase the fraction of students receiving NSE degrees. (2) Finance the creation of a system of objective, achievement-based (rather than normed) tests that measure undergraduate level mastery of various areas of natural science and engineering. (3) Create and fund a new class of portable fellowships, offered to promising young students, that pay $20,000 per year for 3 years of graduate training in natural science and engineering.

The details for all of these programs would have to be adjusted based on more detailed prior analysis and as experience with any of them is acquired in practice. Many alternative programs could also be proposed. These three are offered here primarily to indicate the wide range of possibilities and to move the debate about government programs out of the rut in which it has been stuck for some 20 years.

Training grants could be very flexible. They could follow the pattern that has already been established for training grants at the graduate level. Formally, grants could still be given to a lead principal investigator, but in effect, they would offer financial support to a department at a university or college. The details of the proposed training program would be left open to the applicants. Like all grants, they would be peer reviewed, with fixed terms but renewable. One of the central criteria in evaluating any proposed grant would be some estimates of its cost-effectiveness as measured by the expenditure per additional undergraduate NSE degree granted. At this point, undergraduate institutions in the United States award about 200,000 NSE degrees each year. The vast majority (roughly 95%) of these degrees are awarded to U.S. citizens. It will take an increase of about 100,000 NSE degrees to U.S. citizens per year to meet the goal of having 8% of 24-year-olds receive an NSE degree. If the federal government devoted $1 billion per year, or about $10,000 per additional degree recipient as a reward to schools that could increase the numbers of NSE degrees that they award, universities would surely find it in their interest to reverse the existing pattern of discouraging students from pursuing NSE degrees. Existing liberal arts universities could reallocate resources internally. Specialized science and engineering schools could use these funds to expand. New institutions could enter the educational marketplace.

One of the obvious risks associated with a goal of increasing the number of NSE degrees is the risk that universities would simply relabel existing degrees as NSE degrees or would substantially reduce the content of the NSE degrees that they award. One additional criterion for evaluating training grants would be the presence of metrics that verify whether the quality of the degree from the recipient institution is being eroded. But eventually, it would be more efficient to have objective, national measures of student mastery of science rather than the kind of implicit, idiosyncratic, institution-specific assurances of the quality that universities now provide. The model for this system of measures would be the advanced placement tests offered to high school students by the College Board. This organization has shown that it is possible to construct reliable tests with the property that when teachers teach to the test, the students actually learn the material that they should learn. Just as the AP system is guided by high school and college educators, one would expect that any such system for measuring undergraduate achievement in science would be guided primarily by the professors who teach science at the undergraduate and graduate level. Presumably, scores on these kinds of achievement-based exams would not replace other indicators like course grades, letters of evaluation, and general measures of intellectual ability such as are provided by the existing graduate record exams. Nevertheless, they would provide a new and useful piece of information about performance by individual students, by different educational institutions, and by the nation's educational system as a whole. Given the pervasive problems of incomplete information in higher education, it would surely be of value to students, employers, and faculty members to have access to objective measures of what students actually learn.

The new fellowship program is intended primarily to encourage the process of innovation in graduate education by providing a ready pool of funds that could be spent on any attractive new programs that are created. It would also create additional incentives for students to pursue undergraduate NSE degrees. Possible details for such a program could be as follows: The government could select a sample of graduating high school students who show promise in science, say the more than 100,000 high school students per year who pass the advanced placement exam in calculus. It could offer to a randomly selected treatment subgroup a fellowship that will pay $20,000 per year for 3 years of *graduate* education in natural science or engineering if the student receives an undergraduate NSE degree. (There would be little reason to pay them a subsidy for undergraduate education. Virtually all of these students already go on to get an undergraduate degree.) Granting the award before they begin their undergraduate study would allow them to take the science courses that prepare them for graduate study. Because the treatment group would be randomly selected, it will be easy to verify whether these grants increase the likelihood that a student receives an undergraduate NSE degree. One could also look among the students who continue their studies in graduate school and see whether the recipients of the portable fellowships select career paths that differ from the students who are supported under the existing RA and TA positions. To the extent that fairness is a concern, one could give some other award to the students in the control group, a new personal computer perhaps.

These fellowships would be portable both in the sense that they could be used to pay for training in any field of natural science and engineering and in the sense that they could be used at any institution that the student selects. Some of the students who receive these fellowships would no doubt pursue a traditional course of Ph.D. study, but some may be willing to experiment with other kinds of degrees. Because these funds would represent new funds, not subtractions from the funds that are already used to support graduate students, and because they would only cover 3 years of training, they should not pose much risk to the traditional training system in basic science.

If the government paid for a total of 50,000 of these fellowships each year, or about 16,700 for each annual cohort of students, this would represent an annual expenditure of about $1 billion. (To pay for 16,700 new fellowships each year, the government would presumably have to offer many more because the take-up rate would be less than 100%.) It is possible that the availability of these funds would not lead to the introduction of new courses of study that cater to the recipients. If this were the outcome, the fellowships would be judged a failure and would presumably be discontinued. But, a priori, it seems quite likely that a flow of funds of this magnitude would induce at least some innovative response from our educational system. It should not take many years of observation to verify whether this conjecture is correct.

IX. Conclusions

The analysis here is driven by two basic observations. The first is that better growth policy could have implications for the quality of life in all dimensions that are so large that they are hard to comprehend. The second is that in the last several decades, the efforts that our nation has undertaken to encourage faster growth have been timid and poorly conceived.

We owe it to our children and their children to address questions about growth policy the way we would approach a major threat to public health. We must use the best available evidence and careful logical analysis to frame new initiatives. We must then be willing to run experiments and to see what actually works and what does not.

References

Autor, David, Lawrence F. Katz, and Alan B. Krueger. 1998, November. "Computing Inequality: Have Computers Changed the Labor Market?" *Quarterly Journal of Economics* 113:1169–213.

Frank, Robert, and Phillip Cook. 1995. *The Winner-Take-All Society*. New York: Free Press.

Goolsbee, Austan. 1998, May. "Does Government R&D Policy Mainly Benefit Scientists and Engineers?" *American Economic Review* 88:298–302.

Griliches, Zvi. 1992. "The Search for R&D Spillovers." *Scandinavian Journal of Economics* 94:347–520.

Hall, Bronwyn, and John van Reenen. 1999. "How Effective Are Fiscal Incentives for R&D? A Review of the Evidence." Working Paper no. W7098, National Bureau of Economic Research, Cambridge, MA.

Historical Statistics of the United States: Colonial Times to 1970, Part 1, U.S. Bureau of the Census, Washington, D.C., 1975.

Jones, Charles I. 1995, May. "Time Series Tests of Endogenous Growth Models." *Quarterly Journal of Economics* 110:495–525.

Jones, Charles I., and John C. Williams. 1998, November. "Measuring the Social Return to R&D." *Quarterly Journal of Economics* 113:1119–35.

Katz, Lawrence F., and Kevin M. Murphy. 1992, February. "Changes in Relative Wages, 1963–1987: Supply and Demand Factors." *Quarterly Journal of Economics* 107:35–78.

Langlois, Richard N. and Mowery, David C. 1996, "The Federal Government Role in the Development of the U.S. Software Industry." In David C. Mowery, ed., *The International Computer Software Industry: A Comparative Study of Industry Evolution and Structure.* New York: Oxford University Press.

Lerner, Josh. 1999, July. "The Government as Venture Capitalist: The Long-Run Impact of the SBIR Program." *Journal of Business* 72:285–318.

Maddison, Angus. 1995. *Monitoring the World Economy, 1820–1992*. Paris: Development Centre of the Organisation for Economic Cooperation and Development.

Morgan, Rick, and Len Ramist. 1998, February. "Advanced Placement Students in College: An Investigation of Course Grades at 21 Colleges." Report no. SR-98-13, Educational Testing Service, Princeton, NJ.

National Research Council. 1998. *Trends in the Early Careers of Life Scientists*. Washington, DC: National Academy Press.

National Science Board. 1998. *Science and Engineering Indicators—1998*. Arlington, VA: National Science Foundation.

Rosenberg, Nathan. 1969. *The American System of Manufactures: The Report of the Committee on the Machinery of the United States 1855, and the Special Reports of George Wallis and Joseph Whitworth 1854*. Edinburgh: Edinburgh University Press.

Statistical Abstract of the United States. U.S. Bureau of the Census, Washington, D.C., various years.

Stern, Scott. 1999, October. "Do Scientists Pay to Be Scientists?" Working Paper no. 7410, National Bureau of Economic Research, Cambridge, MA.

Wallsten, Scott. 2000, forthcoming, Spring. "The Effects of Government-Industry R&D Programs on Private R&D: The Case of the Small Business Innovation Research Program." *The Rand Journal of Economics*.